VENUS IN TRANSIT

*

PATHS OF VENUS

(MOST NORTHERLY, CENTRAL, AND MOST SOUTHERLY)

ACROSS THE SUN'S FACE

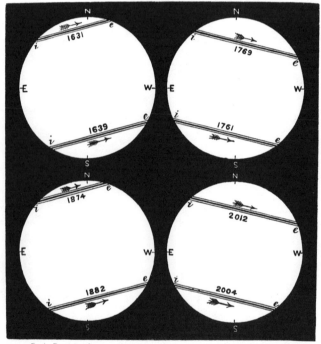

R. A. Proctor ael.

DURING THE TRANSITS OF

A.D. 1631, 1639, 1761, 1769, 1874, 1882, 2004, AND 2012.

VENUS IN TRANSIT

Eli Maor

✳

PRINCETON UNIVERSITY PRESS

PRINCETON AND OXFORD

This paperback is an expanded edition of *June 8, 2004:
Venus in Transit*, originally published in hardcover in
2000 by Princeton University Press

Library of Congress Cataloging-in-Publication Data has
been applied for

ISBN 0-691-11589-3

Excerpts from "Transit of Earth,"
in *The Wind from the Sun: Stories of the Space Age*,
© 1972 by Arthur C. Clarke, reprinted
by permission of Harcourt, Inc.

British Library Cataloging-in-Publication Data is available

This book has been composed in Caledonia

Printed on acid-free paper. ∞

pup.princeton.edu

Printed in the United States of America

10 9 8 7 6 5 4 3 2 1

In memory of my grandparents,

Frida and Karl Stiefel

CONTENTS

PREFACE

CHILDHOOD memories often last a lifetime. When I was four or five years old, my grandmother used to take me out after dinner on Saturday night to watch for the first three stars visible in the evening sky, the moment when, according to Jewish tradition, Shabbat ends. Tel Aviv in those days was a small town, not yet awash in light as most cities are today, and one could still enjoy a fairly clear view of the heavens. I remember my pride when I was able to see those first three stars in the fading glow of dusk! Later she bought me a star map, one of those revolving contraptions where the constellations, etched in white against a dark-blue background, were visible through an oval-shaped window; as you rotated the map and set it to the right day and hour, the entire sky would unfold before you, revealing a miniature image of the universe.

On the night of March 3, 1942, Tel Aviv was treated to the spectacle of a total lunar eclipse. My grandfather prepared me for the event, and with great anticipation the two of us waited for the show to begin. And lo and behold, at the precise predicted moment you could see a small dark notch biting into the full moon. We watched in fascination as Earth's shadow slowly covered the lunar disk, until the entire moon was immersed in darkness, leaving a dim, reddish, ghostly sphere hanging in the sky. What impressed me most, however, was not the event itself but its predictability, the fact that someone could foresee with absolute certainty a heavenly event taking place more than 200,000 miles away—the moon's entrance into earth's shadow.* My grandfather then told me the story of Christopher Columbus, who, when the natives refused to give his men food and water, warned them that God will punish them by taking away the moon's light. His admonition was at first ridiculed, but when at the appointed time the moon's light began to fade, they repented and fell into submission. So, astronomical events not only could be predicted, they could also be put to good use.

Other stories followed: Galileo's defiant exclamation, "But it doeth move!" at the end of his trial by the Roman Inquisition for his belief in the Copernican system, left a deep impression on me. Later I learned that he didn't actually say these words, as the records of the trial show. I also learned that the story about Galileo dropping stones from atop the Leaning Tower of Pisa may not have been true either; and that Newton may never have seen the famous apple fall from the tree. Never mind; these stories were too beautiful to be dismissed!

Out of these early impressions grew a passion for things astronomical. I read every astronomy book I could lay my hands on. The stories of total solar eclipses, with photos showing the black moon completely covering the solar disk, leaving only the pearly corona in sight, made me determined to one day witness this grand spectacle with my own eyes. But the moon's shadow was not to pass over the Holy Land for another hundred years, so I knew I had to go wherever else an eclipse would be visible. My chance came on

*I was able to identify the date of this eclipse with the help of Fred Espenak's *Fifty Year Canon of Lunar Eclipses: 1986–2035* (NASA Reference Publication 1216, March 1989).

February 26, 1979, when I took my wife and our two small children on a two-day trip to Montana. On the night before the eclipse, we drove through a blinding snowstorm, and our hopes dimmed. But Eclipse Day dawned bright and clear, and we saw the event from beginning to end. A spectacular sight it was, worth the long trip.

I was also left with some nonastronomical memories from that event. In describing past eclipses, writers have often commented on the eerie silence during totality, when not only humans but all of nature seems to be awestruck by the grand heavenly spectacle. But during this particular eclipse, someone decided that the excitement of the event itself was not enough: at the instant totality began, the town's sirens went off, as if to announce the arrival of an imminent disaster. I tried to ignore this distraction when I noticed another peculiarity: we were stationed near a major highway, and heavy trucks were rolling by at the crucial moment. I could not imagine how someone could just keep driving and not stop to watch this once-in-a-lifetime experience. But perhaps in the eyes of those truck drivers, *we* were the real oddballs, having traveled 1,500 miles to watch two and a half minutes of darkness.

"The appetite comes with eating," says an old proverb. This first total eclipse created in me an urge to read every piece of eclipse literature I could get hold of. Before I knew it, I became an eclipse addict, willing to spend time and money to chase the moon's shadow whenever and wherever it crossed our planet. Then one day I read in an old astronomy book that on June 8, 2004, the planet Venus will pass in front of the sun, making itself visible as a small black circle slowly moving across the solar disk. A strange eclipse of sorts! But when I searched for literature on this rarest of heavenly events, I found very little, most of it written more than a hundred years ago for the previous transit of Venus. Thus the present book was conceived.

Only five times before have humans witnessed the passage of Venus in front of the sun. The last transit was in 1882, and in our lifetime only two will present themselves. The rarity of the event, combined with the incredible misfortunes that befell some of the scientists who tried to observe it, gives the story all the elements of a true drama. With the layperson in mind, I have limited the use of technical terminology to a bare minimum. I have, however, amply quoted from primary sources to preserve the spirit of the words in their original. I hope that the story told here will recapture some of the excitement that the five past Venus transits have evoked in all those who saw them.

The paperback edition of this book has two new sidebars: one on the recent news that an extrasolar planet has, for the first time, been observed transiting its parent star, and the other on my trip to Manchester, England, and the village of Much Hoole, where Jeremiah Horrocks observed the historic transit of Venus in 1639, the first ever seen. I have also added several new references that became available to me since the first edition was published, resulting in an expanded bibliography.

In the hardcover edition I thanked our son Eyal for designing the line drawings for this book. To our deep sorrow, he passed away on April 26, 2000, after a short battle with cancer. We miss him dearly.

Special thanks go to Helene Auer for translating some of Camille Flammarion's reports of the 1882 transit from the French; to Joe Rao of the Hayden Planetarium and Cablevision's News 12, New York, and to Jack Zirker, former head of the National Optical Astronomical Observatory, for their careful reading of the manuscript and their very useful suggestions; to Fred Espenak of NASA's Goddard Space Flight Center for providing me with updated times for the 2004 transit; to Reny O. Montandon for sending me some hard-to-find publications on past transits of Venus; to Avi Varkowitzky for helping me plan my visit to the village of Much Hoole, England; to the staff of Princeton University Press for their care and dedication in preparing the work for print; to the Skokie and Morton Grove Public Libraries, whose staff greatly helped me in locating rare and out-of-print sources; and last but not least, to my dear wife, Dalia, for her continued encouragement and useful comments as the book was taking shape.

A NOTE ABOUT UNITS

Because much of the literature on transits was written by Englishmen, solar-system distances are given in this book in miles, rather than kilometers. I avoided the common but annoying practice of giving distances in both units, trusting that the reader can make the conversion to kilometers by multiplying by 1.609, a fairly close approximation for most practical purposes.

A NOTE ABOUT SPELLING

There is a tendency among modern authors to capitalize the words for all solar-system objects, including the sun, earth, and moon. I have capitalized these words only when they specifically refer to astronomical objects. For example, "the view from planet Earth," but "a transit of Venus as seen from the earth." When quoting, I have, of course, kept the original spelling.

September 15, 2003

VENUS IN TRANSIT

✳

Prologue

JERUSALEM, Tuesday, June 8, 2004. The sun has just risen over the barren hills of the Judean Desert to the east, where, on a clear day, one can see the Dead Sea and the Moab Mountains beyond. From atop a tall minaret in the Moslem Quarter of the walled Old City, the *muezzin's* monotonous, melancholic chant is already calling the faithful for their morning prayer. In the adjacent Christian Quarter, a lonely church bell chimes in the distance. And in the Jewish Quarter, pious Jews in their black garb solemnly walk down a narrow alley leading to the Western Wall, the sole remnant of King Herod's mighty Temple, where they will pray in devotion, as Jews have done every day for the past two thousand years. In another hour or two, throngs of shoppers and tourists will fill the Old City's narrow alleys, lined up with countless bazaars, small coffeehouses, crafts stores, and souvenir shops. The air will be saturated with the inviting fragrance of spices, the smell of produce and meat, and the aroma of freshly baked bread and *pita*. Another hot summer day in this 3,000-year-old city is about to begin.

But for several hundred people, gathered on the high ridge of Mount Scopus overlooking the city from the east, this will be anything but just another day. Below them to the west, a breathtaking panorama of the city has just unfolded,

dominated by the golden Dome of the Rock glittering in the
first rays of sunlight. From this very ridge, the Roman le-
gions under Titus's command lay siege to Jerusalem in A.D.
68, to be followed by the Crusaders, the Mamluks, the
Turks, the British, the Jordanians, and the Israelis. From
this ridge, countless pilgrims followed Jesus' last journey
down the Via Dolorosa into the walled city; it was here, one
mile to the south, that Jesus was tried and imprisoned.
Kings and statesmen stood in awe at this very place, taking
in the sights and sounds surrounding them. Kaiser Wilhelm
II of Germany came here on a royal visit in 1898; Winston
Churchill, then the British colonial secretary, followed in
1921, and Albert Einstein gave here his inaugural address
at the ceremonies opening the Hebrew University of
Jerusalem in 1923, his first and only speech in Hebrew.

But the group of people gathered here today are directing their gaze not westward but to the rising sun in the east;
their thoughts, at least for now, are not about the 3,000
years of history surrounding them, but on the immediate
future before them. They came here from all over the world
to witness a once-in-a-lifetime heavenly spectacle. Their
cameras and telescopes, carefully covered with protective
filters, are already aimed at the rising sun. Last-minute ad-
justments are made in a hurry, the equipment is checked
and rechecked for any possible glitches, and everyone is
anxiously listening to the latest weather forecast, hoping
and praying that an unexpected cloud will not block the sun
at the last minute. But so far the sky has been clear, as it usually is at this time of the year.

As the minutes tick by, a sense of expectation settles over
the group. Time passes: it is 7:30 in the morning, then 8:00.
Now the tension is almost unbearable. In just a few minutes

these people, like many others at locations as far away as China and Australia, will witness a sight seen by humans only five times before. At precisely 8:19 A.M., a tiny notch, barely visible at first, is seen entering the eastern edge of the solar disk. A loud shout spontaneously erupts from everyone's throat: *first contact!* In the next few minutes the notch slowly encroaches on the sun's face, and the shape of a small black circle clearly makes itself apparent. The spectators are glued to their instruments. Clocks tick, cameras click, and in a few more minutes the black circle will be totally immersed in the sun's disk: *second contact.* It is 8:38 A.M.

The tension now eases a bit, and people excitedly exchange impressions of what they have just witnessed. For the next five and a half hours the black dot—the silhouette of planet Venus projected against the sun—will slowly move across the solar disk from east to west. At 2:04 P.M. third contact occurs, to be followed nineteen minutes later by fourth contact—the instant when the image of Venus finally leaves the sun, not to return until the year 2012.

The passage of Venus in front of the sun is among the rarest of astronomical events, rarer even than the return of Halley's comet every seventy-six years. Only five *transits of Venus,* as the phenomenon is technically called, are known to have been observed by humans before: in 1639, 1761, 1769, 1874, and 1882. But should anyone miss the transit of 2004, all is not lost: the next transit will occur on June 6, 2012, although it will be visible in its entirety only from the Pacific Ocean and the extreme east coasts of Siberia, Japan, and Australia. Then it will be a long wait once again, until December 11, 2117, when Venus will again pass in front of the sun—a bit too far in the future for most of us.

Admittedly, the heavens can offer sights more spectacular than a transit. Nothing can match the grandeur of a total solar eclipse, when for a few precious minutes—and sometimes only seconds—the solar disk is completely covered by the moon, and the pearly corona—the sun's tenuous atmosphere—can be seen glowing around the moon's dark image. But what makes a transit so unique, besides its extreme rarity, is its potential use in determining the value of the *astronomical unit*—the mean distance between the earth and the sun. This distance, about 93 million miles, is known today with great precision, but in the eighteenth century its determination was one of the most daunting challenges facing astronomers, who devised numerous schemes to meet it. To follow the story of these attempts, we must go back to the seventeenth century, when the possibility that Venus may on rare occasions pass in front of the sun was first given serious thought.

1

The Dreamer

The promulgation of Kepler's Laws is a landmark in
history. They were the first "natural laws" in the modern sense: precise, verifiable statements about universal relations governing particular phenomena, expressed in mathematical terms.

Arthur Koestler, *The Watershed* (1960)

IN 1627 JOHANNES KEPLER, mathematician, astronomer,
astrologer and mystic, published his last major work, the
Rudolphine Tables. Dedicated to his patron, Emperor
Rudolph II of Bohemia, it was the most comprehensive
compilation of astronomical data to date; it included rules
and tables for finding the position of the sun, moon, and
planets, a catalog of over one thousand stars begun by his
late mentor Tycho Brahe, improved tables of logarithms,
and the geographical coordinates of major cities of the
world. In the making for over thirty years, the work had
been eagerly awaited by navigators, astronomers, and horoscope casters. Publication was delayed time and again—
first, by the Thirty Year War, then by lack of funds, and
finally by lawsuits from creditors and from Brahe's sons,
who accused Kepler of stealing their late father's observations. The work was finally published in September 1627,
three years before Kepler's death (fig. 1.1).[1]

FIGURE 1.1 Frontispiece of the *Rudolphine Tables*.

Kepler, universally regarded as the founder of modern astronomy, was perhaps the most controversial scientist in history. He was born on December 27, 1571 (by the old Julian calendar then in use) to a family of vagabond misfits in the small town of Weil in the district of Swabia in southwestern Germany. Young Kepler suffered from poor health—real and imagined—and had a very low image of himself. In his diary, which he wrote in the form of a family horoscope, we find this early entry:

> That man has in every way a dog-like nature. His appearance is that of a little lap dog. . . . He liked gnawing bones and dry crusts of bread, and was so greedy that whatever his eyes chanced on he grabbed. His habits were similar. He continually sought the good will of others, was dependent on others for everything, ministered to their wishes . . . and was anxious to get back into their favor. He is bored with conversation, but greets visitors just like a little dog; yet when the least thing is snatched away from him, he flares up and growls. He tenaciously persecutes wrongdoers—that is, he barks at them. He is malicious and bites people with his sarcasm. He hates many people exceedingly and they avoid him, but his masters are fond of him. His recklessness knows no limits . . . yet he takes good care of his life. In this man there are two opposite tendencies: always to regret any wasted time, and always to waste it willingly. . . . Since his caution with money kept him away from play, he often played with himself. His miserliness did not aim at acquiring riches, but at removing his fear of poverty—although, perhaps avarice results from an excess of this fear.[2]

"That man" is Kepler himself, speaking in third person.

Nothing in Kepler's family history showed any hint of a future greatness. His grandfather served as the mayor of Weil, but was, by Kepler's own account, "arrogant, proudly dressed, short-tempered and obstinate . . . his face betrays his licentious past." Kepler's father, Heinrich, one of twelve siblings, was a mercenary who wandered throughout Germany, fighting on the side of whichever religious cause came his way, and narrowly escaping the hangman's rope. Kepler's mother, Katherine, was raised by an aunt who was later burned at the stake for witchcraft, and she herself would barely escape a similar fate late in life. Johannes's opinion of his parents was as harsh as that of himself. His father was "vicious, inflexible, and doomed to a bad end. Saturn in VII [i.e., in Libra, the seventh constellation of the zodiac] made him study gunnery; many enemies, a quarrelsome marriage . . . a vain love of honors, and vain hopes about them; a wanderer . . . 1577 he ran the risk of hanging. . . . Treated my mother extremely ill, went finally into exile and died." His mother was "small, thin, swarthy, gossiping, quarrelsome, and of bad disposition." These are indeed harsh words with which to judge one's parents, and they were matched only by his low opinion of himself. This early flair for self-criticism and brutal honesty would stay with him to the end, and he would use it equally in his personal life and scientific work.

At the age of thirteen he was sent to a theological seminary, where the official language was Latin and strict discipline was the order. Irritable and quarrelsome like his forebears, he made few friends and many enemies; by his own account, he disliked his teachers, and they reciprocated in kind. At seventeen he entered the University of Tübingen, graduating three years later in theology. There he met the one teacher who left a positive impact on him—Michael Mästlin, a professor of astronomy. Through Mästlin he became acquainted with Copernicus's heliocentric (sun-centered) system, and immediately became a fervent believer. But, typically, his beliefs were based on theological rather than sound astronomical reasons: a sun-centered universe made sense to him because God would naturally place the sun, giver of light and heat, at the center of creation. This mixture of true science with religious and mystical reasoning was to be Kepler's hallmark for his entire life.

When he was twenty-three, his life suddenly changed for the better: he was offered a position at the university of Graz in Austria as a teacher of mathematics and astronomy. He accepted, but only reluctantly, citing "the unexpected and lowly nature of the position, and my scant knowledge of this branch of philosophy." At this stage of his life he was still set on a career in theology; but only a year later, while giving his weekly lecture to a nearly empty class, an idea struck him that would remain his credo for the rest of his life.

It happened on July 9, 1595. He was drawing a geometric figure on the board when suddenly a revelation came to him: *God designed the cosmos along simple, geometric proportions.* "The delight that I took in my discovery," he wrote later, "I shall never be able to describe in words." His "discovery"—already expressed two thousand years earlier by the Pythagoreans—was that number and shape are the essence of the universe. But Kepler went further: he proposed that the orbits of the planets around the sun were determined by the geometry of *the five regular solids.* In the two-dimensional plane, one can construct regular polygons with any number of sides—an equilateral triangle, a square, a regular pentagon, etc. (in a regular polygon, all sides are of equal length, and all angles have the same measure); but in space there exist just five regular solids: the tetrahedron, which has four equal faces, each an equilateral triangle; the cube (six faces, each a square); the octahedron (eight equi-

lateral triangles); the dodecahedron (twelve regular pen-
tagons); and the icosahedron (twenty equilateral triangles).
These five solids (fig. 1.2) were already known to the Greeks
and have come to symbolize the perfect symmetry of God's
design; now Kepler made them the cornerstone of his cosmos. "Why are there exactly six planets," he asked, "and
not twenty or a hundred?" His answer: because six planetary orbits leave five gaps between them, and naturally
these five gaps had to be filled with the five regular solids!
It was too good a fit to be a mere coincidence; it had to be
God's design. One almost feels grateful that Kepler did not

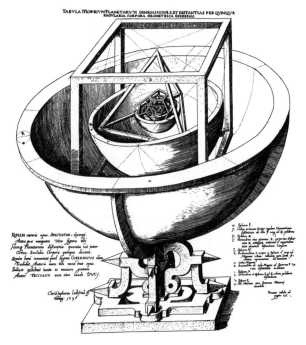

FIGURE 1.2 Kepler's model of the universe, based on the
five regular solids.

know of the three remaining planets of the solar system, for they would have at once destroyed this perfect celestial harmony.

Having made his great "discovery," he now set out to perfect it with a tenacity unparalleled in the annals of science. When the observational data didn't quite fit his vision, he often changed the data—and willingly admitted so later. And when even that didn't quite help, he turned to the laws of musical harmony, assigning to each planet a tune to be sung according to its distance from the sun. Mercury, the closest planet to the sun, was given the highest notes, Saturn the lowest. This "harmony of the spheres" became an idée fixe with Kepler, guiding (perhaps "misguiding" would be a better word) him for the next thirty years. Eventually, aided by the meticulous observations made by the Danish nobleman Tycho Brahe at his elaborate observatory on the island of Hven, he finally discovered the true laws of planetary motion that bear his name. Kepler's three laws are:

1. The planets move around the sun in ellipses, the sun being at one focus of each ellipse.
2. The straight line connecting each planet to the sun sweeps equal areas in equal times.
3. The square of the period of each planet is proportional to the cube (the third power) of its mean distances from the sun.[3]

With these laws, modern astronomy was born.

Kepler was not the first one to ask, *how* do the planets move in their orbits, but he was the first to give the correct answer. By replacing the hallowed circular orbits of the Greeks with elliptical orbits, he discovered the true geometry of the planetary clockwork. Half a century later, Isaac

Newton would use Kepler's laws to answer the *why*—to discover the physical cause that drives this clockwork, the universal force of gravitation.

Notes and Sources

1. For a more detailed description of the *Tables,* see Owen Gingerich, *The Great Copernicus Chase and Other Adventures in Astronomical History* (Cambridge, Mass.: Sky Publishing Corporation, and Cambridge, U.K.: Cambridge University Press, 1992), chapter 15.

2. Kepler's quotations in this chapter are taken from Arthur Koestler's classic, *The Watershed: A Biography of Johannes Kepler* (New York: Anchor Books, 1960).

3. Expressed mathematically, $(T_1/T_2)^2 = (d_1/d_2)^3$, where T_1 and T_2 denote the periods of any two planets, and d_1 and d_2 their mean distances from the sun.

2

Dawn of a New Cosmology

The heavens are immense in comparison with the earth.

Nicolaus Copernicus, *De revolutionibus* (1543)

EVEN IN 1630, the year Kepler died, the Copernican system was by no means universally accepted. On the contrary, most scholars, and more importantly, the Roman Catholic Church, rejected it, the former on physical grounds—after all, no one could really "feel" the motion of the earth—and the latter because the new system ran smack against the teachings of the church's canons. Besides the Bible, these canons included a work of great influence written in the second century A.D. by the Greek geographer-astronomer Claudius Ptolemaeus, commonly known as Ptolemy. Known as the *Almagest*,[1] the work is a summary of the Greek world picture as it was then accepted: a stationary Earth, permanently seated at the center of a finite universe, whose boundary was the celestial dome of the fixed stars. This dome revolved around the earth once in twenty-four hours; in addition, the sun, the moon, and the five known planets (literally, "wanderers") each had its own motion, superimposed on the regular revolution of the celestial dome. It was a reasuringly simple world picture, and it fitted harmoniously with the religious belief that God cre-

ated the universe with man at its center—spiritually as well
as physically.

A few brave men dared to question this world picture,
sometimes with dire consequences. Giordano Bruno, phi-
losopher, dreamer, and preacher, publicly advocated Copernicus's new cosmology, in which the sun replaced the
earth as the center of the world; Bruno went even further,
proclaiming that the sun was but one star in a vast universe,
populated by countless other suns and planets. This was too
much for the Roman Church: Bruno was tried and burned
at the stake in Rome in the year 1600. Galileo Galilei narrowly escaped a similar fate; denounced by the Inquisition,
humiliated by his colleagues, and finally brought to trial for
preaching the new cosmology, he spent his last seven years
under house arrest, dejected and blind.

✳✳✳

ON MAY 24, 1543, when Polish astronomer and clergyman
Nicolaus Copernicus lay on his deathbed, a copy of his
book, *De revolutionibus,* fresh from the printer, was rushed
to him; he was comforted to see his lifelong work come to
fruition just hours before he died. In this great work,
Copernicus demoted Earth from its hallowed position at
the center of the universe, and replaced it with the sun (fig.
2.1). The earth, he asserted, was just another planet, moving in its orbit around the sun like the others. That asser-
tion was revolutionary enough, but he went even further:
the earth, he declared, was but a tiny speck in an immense
universe, vastly bigger than anyone had imagined. These as-
sertions would profoundly change our view of the universe
and our place in it. Yet the popular phrase "the Copernican
Revolution" carries with it a double irony, for the title *De*

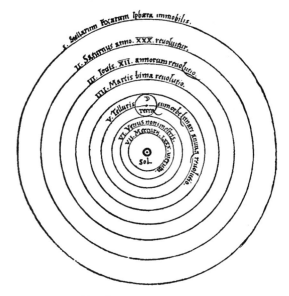

FIGURE 2.1 The Copernican system.

revolutionibus refers not to any political or social revolution, but quite literally to the revolutions of the planets around the sun; and the changes in the world picture he had set out to accomplish were, at least originally, not as revolutionary as is commonly thought.

Copernicus's chief goal was to find a simple explanation to the seemingly complex motion of the heavenly bodies. As every observer of the sky knows, the five planets visible to the unaided eye—Mercury, Venus, Mars, Jupiter, and Saturn—usually move along the zodiac from west to east. But on occasion a planet will appear to slow down, come to a standstill, and then reverse its course to an east-to-west motion, only to reverse it again to its original course. This *retrograde motion* caused the ancients much headache—it

threw into doubt their belief that the planets move around
the earth in perfect circles. The Greeks tried to explain this
anomaly by assuming that a planet actually moves in an
epicycle—a small circle whose center moves along the main
circle around the earth, resulting in a complex, coil-like
curve (fig. 2.2). When even this mechanism didn't quite fit
the observational data, more and more epicycles were
added, until the system got so cumbersome as to become

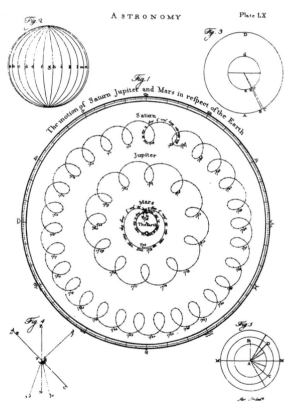

FIGURE 2.2 Epicyles (from a 1798 print).

useless. Copernicus showed that by referring the motion of the planets to the sun, rather than the earth, everything falls neatly into place. In particular, retrograde motion simply follows from the fact that we are watching the planets from an observation platform—our own earth—which is itself moving around the sun; hence a planet will at times appear to move backward, when in fact the earth is simply overtaking it in its orbit, just as a train appears to move backward when your own train is overtaking it. In essence, Copernicus's system, in its original conception, was not much more than a mathematical construct—a far cry from the profound philosophical interpretations that later generations gave it.

Still, Copernicus was well aware of the moral and religious implications of his theory. He compiled his book in six volumes, of which only the first contains an outline of his new theory; the remaining volumes deal with mundane astronomical matters such as spherical trigonometry, the theory of eclipses, and an update of Ptolemy's star catalog. Most of the work was completed in 1533, but Copernicus was reluctant to publish it, no doubt fearing it might provoke the Catholic Church. It was only after persistent prodding by his few disciples that he finally relented, but the publication process started slowly and was interrupted time and again. When the galleys were finally brought to him, he was already an old and sick man, barely able to correct the proofs. It thus happened that a preface, written not by Copernicus but by his editor, Andreas Osiander, made its way into the final version. In this preface Osiander (1498– 1552), a Lutheran minister who was active in the reform movement, in effect denied much of what Copernicus said in the book. Osiander probably wanted to protect his own skin from accusations of heresy, for Martin Luther himself was firmly against the new theory. But whatever his intentions, the damage was done: Copernicus's reputation as a scrupulous fighter for the truth was tarnished, since it was assumed that the preface was his own. It was only in 1609 that the truth came out, when Johannes Kepler discovered in a copy of *De revolutionibus* a note that identified the true author of the preface. It was not the first time, nor the last, that an accused astronomer would be posthumously vindicated by one of his followers.

✳✳✳

THE first concrete evidence that the heliocentric system was better suited to describe the planetary orbits came in the epochal year 1609, when Galileo aimed his telescope (he called it "spyglass") skyward. What he saw changed the course of astronomy. The sun, instead of being the perfect, immaculate disk that the ancients had imagined, was peppered with dark spots whose number and position changed from day to day. The moon revealed itself to be a world of mountains, craters, and flat plains. Venus showed phases like the moon—a narrow crescent at one time, a gibbous shape at another, and occasionally a nearly full disk. Here was solid proof that Venus moved around the sun; for had it moved around the earth instead, it would have shown a full disk at each opposition, when it was directly opposite to the sun. Then on January 7, 1610—we know the date because he kept meticulous notes of his observations— Galileo turned his telescope to Jupiter and saw four small companions orbiting the giant planet, a solar system in miniature. This clearly showed that small bodies move around large ones, and it pulled the rug under those who still believed that the entire universe revolves around our earth.

By the beginning of the seventeenth century, then, the shortcomings of the old Ptolemaic world picture were there for all to see. But it was not until 1687, with the publication of Isaac Newton's great masterpiece, the *Principia,* that the old system was given its death knell.[2] Newton realized that a falling apple and a planet orbiting the sun are subject to one and the same law—the universal law of gravitation. He was thus the first great unifier in the history of science. The Newtonian universe would dominate our scientific thought for the next 250 years. It would celebrate one triumph after another—from the accurate prediction of comet Halley's return on Christmas Day, 1758, to the discovery of a new planet, Neptune, a century later—and it turned the last remaining skeptics into fervent believers.

Yet there was one remaining unknown in this picture. Kepler's three laws gave astronomers a precise, quantitative theory on which to base their planetary calculations. In particular, his third law (see page 13) allowed them to compute the distance of one planet from the sun in terms of that of another. But it was still a *comparison* of distances, not an absolute determination. It is as if you were studying a road map that has no scale on it: you could tell that a certain town is twice as far from your location as another town, but you couldn't say how many miles it is to each town. If the distance from the sun of just one planet could somehow be found, those of all other planets could at once be computed from Kepler's third law, and the dimensions of the solar system would be known. Naturally, the most obvious planet from which to start was our own. To determine Earth's distance from the sun—the length of the *astronomical unit*—

thus became one of the most pressing challenges facing seventeenth-century astronomy. Chances of success looked dim, but in November 1631 an event occurred that gave scientists a glimmer of hope.

Notes and Sources

1. The original title was *Syntaxis mathematica* (Mathematical compendium), to which later generations added the superlative *magiste* ("greatest"), to distinguish it from other, less important works. When Moslem scholars translated the work into Arabic, they added the prefix "al" ("the"), and soon the work became known in Europe as the *Almagest*. The first Latin translation appeared in 1175, and from then on until the sixteenth century it dominated the astronomical thinking of Europe.

2. The full title is *Philosophiae naturalis principia mathematica* (Mathematical principles of natural philosophy). The universal law of gravitation says that any two bodies in the universe attract each other with a force proportional to their masses and inversely proportional to the square of the distance between them; mathematically expressed, $F = km_1m_2/d^2$, where k is a universal constant.

3

A Sight Never Seen Before

> I have been more fortunate than those hunters after
> Mercury who sought the cunning god in the sun. I
> found him out, and saw him where no one else had
> hitherto seen him.
>
> Pierre Gassendi (French scientist, 1592–1655)

THE *Rudolphine Tables* was Kepler's last major work. He spent his remaining years compiling annual calendars, casting horoscopes for the nobility, and computing future astronomical events, all based on his tables. He also completed a work of fiction begun many years earlier, *Somnium,* in which he described an imaginary trip to the moon. This work—some of the characters are allusions to people who played a role in his own life, including Tycho Brahe—came remarkably close to describing modern space travel; it was published in 1634, four years after Kepler's death.[1] By and large, however, his lifework was done.

As he was computing the dates of some future celestial conjunctions, Kepler made two startling predictions: on November 7, 1631, the planet Mercury would pass directly in front of the sun, and exactly a month later, on December 6, Venus would do the same. Now, each of these celestial rendezvous is a rare enough event in itself, but their occurrence so close to each other made them a truly unique happening. Aware of the potential significance of these events for determining the size of the two planets, Kepler and his assistant, Jacob Bartsch (soon to become Kepler's son-in-law), issued an "admonition" to all living astronomers to be on the watch. Not fully trusting the accuracy of his own tables, he urged them to begin their vigil a day early, and, should they see nothing, not give up until the day after.

Kepler's calculations showed that the transit of Venus on December 6 would not be visible in Europe, so he extended his request to mariners who might be sailing the oceans on that day, as well as to learned men in the New World. As far as is known, no one witnessed this rare event (it actually took place during the night between December 6 and 7, European time). The transit of Mercury on November 7, on the other hand, was visible from Europe, but due to bad weather only three individuals actually observed it. Of them, only one, Pierre Gassendi (1592–1655), left us with a detailed account.

Gassendi was born in Provence, France, and studied at the University of Aix. Being of an independent mind, he rebelled against prevailing medieval doctrines that based science on philosophical rather than physical principles. In this he was strongly influenced by Galileo's experimental approach to science. A major debate in those days was whether an object in free fall shares earth's rotation under it; to prove that it does, Gassendi climbed atop the mast of a boat and dropped a stone while the boat was moving. As expected, the stone landed exactly at the bottom of the mast, proving that even as it was falling, the stone was sharing the ship's forward motion.

Like most scholars of his time, Gassendi's work covered

a wide range of subjects; he was an early advocate of the atomic theory that regarded matter as made of tiny, indivisible units. In 1621 he observed a display of northern lights and gave it the name "aurora borealis." He studied sound waves and showed that they propagate at a speed that is independent of their pitch, in contradiction to the teaching of Aristotle. He also wrote the biographies of several famous scientists, among them Copernicus, Tycho Brahe, and the Renaissance mathematician Johann Müller, known by his Latinized name, Regiomontanus. But Gassendi is mainly remembered today for his account of the transit of Mercury, the first ever to be seen by humans.

As the crucial day approached, Gassendi was ready at his telescope in his apartment in Paris. His plan was to project the sun's image onto a white screen placed at some distance behind the telescope's eyepiece.[2] This was a significant improvement over the *camera obscura* method, favored by Kepler, in which the sun's rays passed directly through a narrow slit into a darkened room, forming an inverted image on the opposite wall. The weather was not cooperating: on November 5 it rained all day, and on the sixth—the day before the predicted transit—it was still hopelessly clouded. Gassendi feared that the transit might have already taken place; but on the morning of the seventh there was a partial clearing, and at 9 A.M. the sun peeked briefly through the clouds. On his screen there appeared the solar image, crisp and clear, about eight inches in diameter. The usual sunspots where there, but he also saw a tiny black dot that had not been there when he observed the sun a few days earlier. Just then clouds intervened once more, and his heart sank. But when the sun came out again, the small dot had changed its position relative to the other spots, and Gassendi realized it must be the image of Mercury on the sun. It was much smaller than he had expected, and for a while he was doubtful whether it wasn't a sunspot after all; but when the sun once more peeked through the clouds, the dot had changed its position again, and Gassendi was finally convinced that it was, indeed, his prized object.

Gassendi was anxious to determine the exact times of Mercury's entrance into and exit from the sun's disk—the moments of *ingress* and *egress*. Precise clocks were not yet available in his day, so his only option was to measure the sun's altitude above the horizon and use it as an indication of the time. Since his own room was darkened for the event, he had an assistant stationed on the floor above his own, ready with a large quadrant (an instrument for measuring celestial angles), with instructions to take the sun's altitude each time Gassendi would stamp his foot on the floor. Of what happened next, we have an amusing report:

> It was clear that the phenomenon which had been so long and so anxiously awaited by the astronomer was already in progress. He immediately stamped upon the floor to attract the notice of his assistant. But this person, whose name has not reached us, was possessed of less patience than Gassendi. He probably felt much less interested in the phenomenon; possibly, he placed very little faith in Kepler's calculations. Whatever the reason, he had grown weary of watching, and had left his post. Gassendi had to continue his observations alone, hoping that at least his assistant would return before the planet had passed completely off the sun's face. Fortunately this happened; the

requisite observations were made for determining the time of egress, and an important addition was made to our knowledge of the motions of the innermost planet of the solar system.[3]

Gassendi's chief goal was to make a rough estimate of Mercury's apparent size. On his screen he drew a circle with two diameters at right angles to each other, each divided into sixty parts. He then adjusted the screen's distance from the eyepiece until the sun's image filled the entire circle. Since the sun's apparent diameter in the sky is about 30 minutes of arc (30′), each division marked half a minute. With this primitive device, he estimated Mercury's apparent diameter to be a mere 20 arc *seconds* (20″); this is equivalent to the angle a dime would subtend when viewed from a distance of 200 yards.

When Gassendi reported his finding, it was received with much skepticism, even disbelief. Nothing thus far had given astronomers the faintest clue as to the size of the planets; their estimates had all been based on pure speculation. Kepler's third law allows one to compute the relative *distances* of the planets from the sun, but the law says nothing about their relative *sizes*, let alone actual dimensions. Kepler's lifelong obsession with harmonic proportions had led him to propose that the *volumes* of the planets are proportional to their distances from the sun. But this would make the inner planets Mercury and Venus much too big, and the outer planets Jupiter and Saturn much too small, compared with their actual sizes. In addition, the brilliance of Mercury and Venus in the sky can easily cause one to exaggerate their apparent size, as Galileo was already aware. Thus it is no wonder that Gassendi's announcement came as a surprise to astronomers, who devised various "reasons" to explain it, none of them convincing. Slowly it dawned on them that size and distance in the solar system are entirely unrelated quantities.

Kepler himself did not take part in this debate, nor did he witness Mercury's passage in front of the sun: he died on November 15, 1630, almost a year to the day before the event he had predicted with such accuracy. Indeed, the transit took place within five hours of the predicted moment—an astonishing feat for his time. In a letter to Kepler's old friend, Wilhelm Schickard, the exuberant Gassendi wrote: "But Apollo [the Roman god of sunlight], acquainted with [Mercury's] knavish tricks from his infancy, would not allow him to pass altogether unnoticed. . . . I have been more fortunate than those hunters after Mercury who have sought the cunning god in the sun. I found him out and saw him where no one else had hitherto seen him."[4] He was referring to the difficulty of seeing Mercury even under normal circumstances: it never wanders more than 28 degrees from the sun, and is usually drowned in the sun's glare, even when at its brightest. Copernicus is said to have lamented that he had never seen Mercury, a claim that is entirely believable, considering the high latitude of his native Poland, where twilight is particularly long. Galileo, using his telescope and observing from the more moderate latitudes of Italy, did see it, but could not discern its disk. Even to a modern-day amateur, finding Mercury can be a challenge— except during a total solar eclipse, when for a few short minutes the sun's glare is hidden by the moon.[5] Gassendi was the first to see Mercury in its true shape, "stripped naked and starkly outlined against the sun's disk."[6]

Notes and Sources

1. See *Kepler's Somnium: The Dream, or Posthumous Work on Lunar Astronomy,* translated with a commentary by Edward Rosen (Madison, Wis.: University of Wisconsin Press, 1967).

2. Albert van Helden, *Measuring the Universe: Cosmic Dimensions from Aristarchus to Halley* (Chicago: University of Chicago Press, 1985), pp. 97–98. A somewhat different account, not mentioning the telescope, is given by Richard Anthony Proctor in his classic book, *Transits of Venus: A Popular Account of Past and Coming Transits* (1874; 4th ed. London: Longmans, Green, and Co., 1882), p. 5. Proctor (1837–1888) was born in Cambridge, England, and became a prolific author of popular works on astronomy. He seems to have been the first to suggest that the lunar craters originated from intense meteoritic bombardment during the early days of the solar system; this view was in contrast to the volcanic theory then prevailing.

3. Proctor, *Transits of Venus,* pp. 8–9.

4. Owen Gingerich, *The Great Copernicus Chase and Other Adventures in Astronomical History* (Cambridge, Mass.: Sky Publishing Corporation, and Cambridge, U.K.: Cambridge University Press, 1992), p. 131.

5. During the total eclipse of February 26, 1998, Mercury was easily visible near the sun, its brightness almost matching that of Jupiter on the opposite side of the eclipsed sun. It was a spectacular sight which I was fortunate to observe from the island of Curaçao.

6. Van Helden, *Measuring the Universe,* p. 99. For a more detailed history of early Mercury transits, see Gingerich, *The Great Copernicus Chase,* chapter 6. See also Jean Meeus, *More Mathematical Astronomy Morsels* (Richmond, Va.: Willmann-Bell, 2002), chapter 48.

4

Venus Stripped Bare

> I hope to be excused for not informing other of my friends of the expected phenomenon; but most of them care little for trifles of this kind, preferring rather their hawks and hounds, to say no worse; and although England is not without votaries of astronomy, with some of whom I am acquainted, I was unable to convey to them the agreeable tidings, having myself had so little notice.
>
> Jeremiah Horrocks
> (English astronomer, 1618–1641)

As far as is known, the transit of Mercury on November 7, 1631, was the first such event ever to be observed by humans. Gassendi, one of the three people known to have watched it, was now determined to observe the transit of Venus, scheduled for just one month later, on December 6. He set up his equipment as he had done for Mercury, and started his watch on December 4. This time luck was not with him:

> An impetuous storm of wind and rain rendered the face of the heavens invisible on both days [December 4 and 5]. On the 6th he continued to obtain occasional glimpses of the sun till a little past three o'clock in the afternoon, but no indication of the planet could be discerned upon

the sun's disk as depicted upon the white circle. On the 7th he saw the sun during the whole forenoon, but he looked in vain for any trace of the planet."[1]

We know today that the transit actually took place on the night between the 6th and 7th of December and was visible in the Western Hemisphere, but there is no record that anyone actually saw it.[2]

Kepler's calculations showed that Venus would not pass again in front of the sun until June 6, 1761. But when an obscure young English astronomer, Jeremiah Horrocks (or Horrox, as his name was then spelled; 1619–1641), examined Kepler's tables, he realized that, unnoticed by Kepler, a transit of Venus would occur on December 4, 1639—just eight years hence! This was a startling discovery, and it showed that even the *Rudolphine Tables* were not entirely free of error.

Horrocks was born to a poor family who lived not far from Liverpool, but his exact place and date of birth are unknown. He showed an early interest in mathematics and astronomy and attended Cambridge University without, however, graduating. In 1639 he moved to the small hamlet of Hoole (now known as Much Hoole), 15 miles north of Liverpool, to accept a teaching and clerical position with the local church. But he lived on a "very poor pittance" and must have disliked his teaching, which was described as "daily harassing duties." His real passion was astronomy, which he studied on his own. He was one of the first Englishmen to fathom Kepler's new planetary theory, and soon became an ardent believer. When only nineteen years old, he showed that the moon's orbit around the earth is an ellipse, with the earth at one focus—this at a time when even Galileo was still holding to the old belief that the planets and their satellites move around their parent bodies in perfect circles. Horrocks repeatedly tested Kepler's third law, relating the periods of the planets to their mean distances from the sun, and found it to be absolutely true. This convinced him that Kepler's theory—which was by no means universally accepted yet—was superior to all others.

In 1635, the sixteen-year-old Horrocks began to compute astronomical ephemerides—future positions of the sun, moon and planets—based on the tables of a Belgian astronomer, Philip van Lansberge (1561–1632), which were more recent than Kepler's. Lansberge had boasted that his tables were superior to Kepler's, but Horrocks found this to be a vain boast: he discovered numerous errors in Lansberge's calculations, putting into question the very method on which their author had based his work. But Lansberge's tables did show that Venus would pass in front of the sun on December 4, 1639. Horrocks, confronted with the two conflicting sets of data, resolved to do his own calculations. After three years of hard work he concluded that Venus would, indeed, be visible on the sun's face on that day.

It wasn't that Kepler was totally ignorant of the circumstances. His tables showed that on December 4 Venus would indeed pass between the earth and the sun, but it would do so just below the sun's disk, so that a transit would not occur. The reason for this narrow miss was that Kepler's tables were initially computed for an imaginary observer situated at the center of the earth; but because we live on the surface of the earth, the sun would appear to us slightly shifted in the sky relative to its position as seen from the center. This is the familiar phenomenon of *parallax*. Stretch your hand to its full length, raise your thumb, and watch its position first with your right eye, then with your left. The thumb will seem to change its position slightly relative to the remote background. The closer your thumb is to your eyes, the greater its apparent change in position. The phenomenon of parallax was known already in the second century B.C. to the Greek astronomer Hipparchus, who used it to estimate the distance between the earth and the moon (see page 44).

Kepler, of course, was thoroughly familiar with the parallax effect, and he corrected his figures to take it into account. But, not knowing the distance between the earth and the sun, he had no way of estimating the actual angle of parallax. In fact, he greatly underestimated the sun's distance (his value was 3,500 earth radii, compared to the true value of about 25,000), leading to an *overestimated* parallax value. He thus concluded that Venus, *as seen from earth's surface,* would just miss the sun on December 4, 1639. When Horrocks recalculated Venus's position, he found that in this case Lansberge was right: Venus would indeed transit the sun on that day.[3]

***

HORROCKS completed his calculations in October 1639, barely a month before the scheduled transit. He immediately alerted some of his friends, exhorting them to observe the rare event from different locations, so as to "less likely be defeated by the accidental interposition of clouds, or any fortuitous impediment." One "fortuitous impediment" he was worried about was that "Jupiter and Mercury seemed by their positions to threaten bad weather; for, in such apprehension I coincide with the opinion of the astrologers, because it is confirmed by experience; but in other respects

I cannot help despising their puerile vanities." Astronomy
and astrology, in those days, were still on good terms, often
practiced by the same person.

Among the astronomers Horrocks had alerted was a
young friend, William Crabtree, a linen draper living in
Manchester, "a person who has few superiors in mathematical learning." The two had met at Cambridge and
shared a passion for mathematics and astronomy; after leaving the university they maintained their friendship only by
correspondence, even though they lived just 25 miles apart.
On November 5 Horrocks wrote to him:

> The reason why I am writing to you now is to inform you
> of the extraordinary conjunction of the Sun and Venus
> which will occur on November 24 [by the "old style" or
> Julian calendar]. At which time Venus will pass across the
> Sun. Which, indeed, has never happened for many years
> in the past nor will happen again in this century. I be-
> seech you, therefore, with all my strength, to attend to it
> diligently with a telescope and to make whatever obser-
> vation you can, especially about the diameter of Venus,
> which, indeed, is 7' [7 arc minutes] according to Kepler,
> 11' according to Lansberge, and scarcely more than 1'
> according to my proportion.[4]

Clearly, Horrocks's main interest in the transit was the opportunity it afforded him to measure Venus's apparent diameter—a task almost impossible to achieve at any other
time, because of the planet's intense glare. The possibility
of using the transit to estimate Venus's *distance* had not occurred to him—or if it had, he didn't tell us.

Horrocks now prepared himself for the upcoming event
as thoroughly as Gassendi had done eight years before him.

Indeed, Gassendi was his role model, and Horrocks lavished praise on him:

> Following the example of Gassendi, I have drawn up an account of this extraordinary sight, trusting that it will not prove less pleasing to astronomers to contemplate Venus than Mercury, though she be wrapt in the close embrace of the sun. . . . Hail! then, ye eyes that penetrate the inmost recesses of the heavens, and, gazing upon the bosom of the sun with your sight-assisting tube, have dared to point out the spots on that eternal luminary! And thou, too, illustrious Gassendi above all others, hail! thou, who first and only, didst depict Hermes' changeful orb in hidden congress with the sun. Well hast thou restored the fallen credit of our ancestors, and triumphed o'er the inconstant wanderer. . . . Contemplate this most extraordinary phenomenon, never in our time to be seen again! The planet Venus, drawn from her seclusion, modestly delineating on the sun, without disguise, her real magnitude, whilst her disc, at other times so lovely, is here obscured in melancholy gloom; in short, constrained to reveal to us those important truths, which Mercury on a former occasion confided to thee.

"Hermes" was the mythological Mercury of the Greeks, and the "inconstant wanderer" was a reference to the difficulties of observing this innermost planet, due to its proximity to the sun.

Horrocks set up his equipment—a small telescope he had bought the year before for half a crown—at his home in Hoole, and anxiously awaited the crucial day. According to his calculations, Venus would enter the sun's disk at 3:57 in the afternoon on December 4; but just in case he might have erred, he began his vigil already on December 3, not willing to take any chance. December 4 dawned; it was a Sunday. Horrocks was at his telescope from sunrise until nine, and again from ten to noon, intently observing the sun's image on the screen in his darkened room. The sky was overcast, but Horrocks was able to get a glimpse of the sun during brief breaks in the clouds. But except for a few sunspots, he saw nothing unusual. At one in the afternoon, his vigil was suddenly interrupted by "business of the highest importance, which for these ornamental pursuits I could not with propriety neglect." What this business of highest importance was he did not say, but from a brief biography of Horrocks, published more than two hundred years later by the Reverend Arundell Blount Whatton, we know that on that Sunday he had the official duty of conducting divine services at his church.[5]

When Horrocks resumed his watch at fifteen minutes past three,

> The clouds, as if by Divine interposition, were entirely dispersed, and I was once more invited to the grateful task of repeating my observations. I then beheld a most agreeable spectacle, the object of my sanguine wishes, a spot of unusual magnitude and of a perfectly circular shape, which had already fully entered upon the sun's disc on the left, so that the limbs of the sun and Venus precisely coincided, forming an angle of contact. Not doubting that this was really the shadow of the planet, I immediately applied myself sedulously to observe it.

Just as Gassendi had missed taking the sun's altitude at the beginning of Mercury's transit because of his assistant's tardiness, so did Horrocks miss the entrance of Venus on the sun because of divine duties! This missed opportunity was, in the words of the astronomer Simon Newcomb, "a circumstance which science has mourned for a century past, and will have reason to mourn for a century to come."[6]

Horrocks, however, did not waste time lamenting over what he had missed; he immediately took a series of measurements of the planet's apparent size and direction of motion across the solar disk (fig. 4.1). He found Venus's apparent diameter to be about one arc minute, again (as with Mercury) much smaller than anyone had thought. He had intended to record every detail of the rare apparition, fully aware that he was making history; alas, it was December, and the days were at their shortest. At 3:50 in the afternoon the sun set, and his observation came to an end. He had been watching the historic transit for scarcely half an hour.[7]

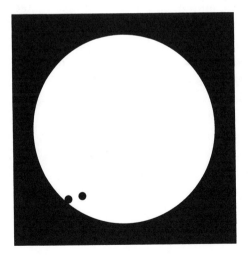

FIGURE 4.1 The transit of December, 4, 1639 as seen by Horrocks.

Horrocks wrote up a detailed report of what he saw; it ended in these words:

This observation was made in an obscure village, where I have long been in the habit of observing, about fifteen miles to the north of Liverpool, the latitude of which I believe to be 52° 20′, although by the common maps it is stated to be 54° 12′; therefore the latitude of the village will be 53° 35′, and the longitude of both 22° 30′ from the Fortunate Islands, now called the Canaries. This is 14° 15′ to the west of Uraniburg, in Denmark, the longitude of which is stated by [Tycho] Brahe, a native of the place, to be 36° 45′ from these islands.

This is all I could observe respecting this celebrated conjunction during the short time the sun remained in the horizon: for although Venus continued on his disk for several hours, she was not visible to me for longer than half an hour, on account of his so quickly setting. Nevertheless, all the observations which could possibly be made in so short a time I was enabled by Divine Providence to complete so effectually that I could scarcely have wished for a more extended period. The inclination [of Venus's path across the sun] was the only point upon which I failed to attain the utmost precision; for, owing to the rapid motion of the sun, it was difficult to observe with certainty to a single degree; and I frankly confess that I neither did nor could ascertain it. But all the rest is sufficiently accurate, and as exact as I could desire.

Even at a distance of 350 years, one is moved by the modesty and honesty of these words.

✱✱✱

HORROCKS was not the only one to witness the transit of
1639. His friend William Crabtree was ready to watch the
event from his home in Manchester, 25 miles east of Horrocks's location, but was beset by cloudy skies during most
of the transit. He was just about to give up his watch, when
at 3:35 in the afternoon the sun suddenly burst out of the
clouds. "He at once began to observe, and was gratified by
beholding the pleasing spectacle of Venus upon the sun's
disc." But, Horrocks tells us, Crabtree was so awestruck by
the sight before him that for a few moments he stood there
motionless, overcome with emotion. Horrocks added, almost apologetically:

> For we astronomers have, as it were, a womanish dispo-
> sition, and are overjoyed with trifles, and such small mat-
> ters as scarcely make an impression upon others; a sus-
> ceptibility which those who will may deride with
> impunity, even in my own presence; and if it gratify
> them, I too will join in the merriment. One thing I re-
> quest: let no severe Cato [Marcus Porcius, Roman
> statesman, known as "the Censor"] be seriously offended
> with our follies; for, to speak poetically, what young man
> on earth would not, like ourselves, fondly admire Venus
> in conjunction with the sun, what youth would not dwell
> with rapture upon the fair and beautiful face of a lady,
> whose charms derive an additional grace from her for-
> tune?

By the time Crabtree regained his senses, the sun had almost set, and he hastily made a few measurements; these

proved to be consistent with those of Horrocks, to the latter's great delight.

The historic transit was now over. Horrocks was aware that he and Crabtree were probably the only humans who saw the rare event, but he added: "If others, without being warned by me, have witnessed the transit, I shall not envy their good fortune but rather rejoice, and congratulate them on their diligence. Nor will I withhold my praise from anyone who may hereafter confirm my observations by their own, or correct them by anything more exact." He concluded his report in words more befitting a poet than a scientist:

> Venus was visible on the sun throughout nearly the whole of Italy, France, and Spain; but in none of those countries during the entire continuance of the transit.
>
> But America! Venus! what riches dost thou squander on unworthy regions which attempt to repay such favours with gold, the paltry product of their mines. Let these barbarians keep their precious metals to themselves, the incentives to evil which we are content to do without. These rude people would indeed ask from us too much should they deprive us of all those celestial riches, the use of which they are not able to comprehend. But let us cease this complaint, O Venus! and attend to thee ere thou dost depart.

It was time to say good-bye to Venus, and Horrocks gave her a rousing adieu:

> Oh! then farewell, thou beauteous queen!
> Thy sway may soften natures yet untamed,

Whose breasts, bereft of native fury,
Then shall learn the milder virtues.
We, with anxious mind, follow thy latest footsteps
    here,
And far as thought can carry us;
My labours now bedeck the monument for future
    times
Which thou at parting left us. Thy return
Posterity shall witness; years must roll away,
But then at length the splendid sight
Again shall greet our distant children's eyes.

He was about to finish writing up his account of the transit, entitled *Venus in sole visa* (Venus visible on the sun), when he suddenly died on January 3, 1641, not yet twenty-two years old; the cause of death is not known. His friend Crabtree, whom Horrocks was to meet the next day—it would have been their first meeting since their student years—survived him by only three years; he reportedly was killed in 1644 in the battle of Naseby field. Were it not for Crabtree, who kept many of Horrocks's letters, we may have never heard of Horrocks.[8] A tribute paid to Horrocks many years later described him as "a prodigy for his skill in astronomy; had he lived, in all probability he would have proved the greatest man in the whole world in his profession." The house in Hoole where he lived and from which he observed the transit—known as the Carr House—still exists today, as does St. Michael's Church, where Horrocks held the service on that famous Sunday. And Crabtree's observation of the historic transit was immortalized by the artist Ford Madox Brown (1821–1893) in a large mural that adorns the town hall of Manchester (plate 8).[9]

As far as is known, Horrocks and Crabtree were the only humans to witness the rare heavenly spectacle. But when Venus appeared again on the sun's face in 1761, the entire astronomical community was awaiting to greet her.

Notes and Sources

1. Richard A. Proctor, *Transits of Venus: A Popular Account of Past and Coming Transits* (1874; 4th ed. London: Longmans, Green, and Co., 1882), p. 11. Unless noted otherwise, all quotations in this chapter are from Proctor.

2. Proctor recalculated the circumstances of this transit and found that on the morning of December 7, the egress was just visible after sunrise from the southeastern parts of Europe. A map in his book shows Italy just on the borderline of the area of visibility. This raises the possibility that perhaps somebody did witness this transit. Galileo, who was in correspondence with Kepler, no doubt knew about the upcoming event, but there is no record that he saw it.

3. In his *A History of Astronomy from Thales to Kepler* (1906; rpt. New York: Dover, 1953), p. 420, the historian J.L.E. Dreyer claims that "they [Lansberge's tables, published in 1632] probably owed a great deal of the good repute they enjoyed for some time to the circumstance that they by a fluke represented the transit of Venus in 1639 fairly well, while the *Rudolphine Tables* threw Venus quite off the sun's disk." Today Lansberge's name is all but forgotten.

4. Albert van Helden, *Measuring the Universe: Cosmic Dimensions from Aristarchus to Halley* (Chicago and London: University of Chicago Press, 1985), p. 107.

5. This, however, has recently been disputed. According to historian of astronomy Allan Chapman, the twenty-year-old Horrocks would have been too young to qualify for conducting official church services. His role that afternoon may have been a minor one, or he may have simply taken the children under his care to the church. Whatever his actual duties that afternoon, they became part of popular science lore. See

Allan Chapman, *Jeremiah Horrocks and Much Hoole* (n.p.: Allan Chapman, 1994), pp. 3, 7.

6. Simon Newcomb, *Popular Astronomy* (New York: Harper Brothers, 1880), p. 178.

7. To quote Chapman, *Jeremiah Horrocks and Much Hoole*, p. 7: "That half-hour was one of the most momentous moments in British astronomy. . . . By knowing exactly where Venus was, with relation to the Sun, at 3:45 P.M. on November 24th [old style], 1639, Horrocks was able to make considerable advances in our knowledge of planetary motion. It was the first major achievement of British astronomical research, and had international repercussions. The scientific conclusions that Jeremiah correctly derived from the observation provided dramatic confirmation of the work of Kepler, Tycho, Galileo and Copernicus."

8. Many of Horrocks's papers were destroyed in the English civil war; others went up in flames in the Great fire of London. Of the papers that remained, many were taken by his brother and never returned. Fortunately, Horrocks's letters to Crabtree were bought by an antiquarian dealer and thus survived. His *Venus in sole visa* was published by the German astronomer Johannes Hevelius in 1662.

9. The visitor who wishes to see these places may consult Charles Tanford and Jacqueline Reynolds, *The Scientific Traveler* (New York: John Wiley and Sons, 1992), pp. 62–63 and 125–126.

Solar and Stellar Parallax

THE METHOD of parallax is one of astronomy's most efficient tools in measuring the distances of celestial objects. It was first used by Hipparchus of Nicaea in the second century B.C. to estimate the moon's distance. On March 14, 189 B.C., a solar eclipse occurred not far from his birthplace. This eclipse was total near the Hellespont (the straight of Dardanelles in modern Turkey), whereas in Alexandria (Egypt), only four-fifths of the sun's disk was hidden by the moon. Since the sun and moon subtend about half a degree of arc in the sky, the moon's apparent shift in position amounted to one-fifth of this, or about 6 arc minutes. Based on this information and on the known distance between the two locations, Hipparchus was able to estimate the moon's distance as between 71 and 83 earth radii. While these estimates are in excess of the modern values of 56 and 64, they came within the correct order of magnitude and were a remarkable achievement for his time.

To use the method of parallax, we choose a *baseline* (line *AB* in fig. 1) whose length is known. Observers at *A* and *B* measure the angles between this baseline and the lines of sight to a distant object *C*. This gives us a triangle *ABC* in which one side and two angles are known; using elementary trigonometry, we can then

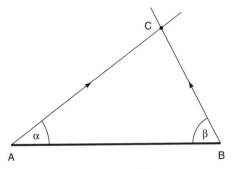

FIGURE 1 Terrestrial parallax.

compute the length of sides *AC* and *BC*, that is, the distance from each observer to the object. This procedure is routinely used in terrestrial surveying. In astronomy, however, *AC* and *BC* are so much longer than the baseline that we can safely take them as equal in length; in other words, *ABC* is now an isosceles triangle (fig. 2). In this case the observers at *A* and *B* only need to measure the shift in the direction of *C* (that is, angle γ) on the celestial sphere; a simple calculation then gives the distance from *A* and *B* to *C*.

Clearly, the longer the baseline, the larger the apparent shift in position of the object, and the greater the accuracy in determining its distance. For a relatively nearby object such as the moon, a short baseline—say a few hundred miles—would suffice. But to find the distances to more remote objects—the sun, for example—we need the entire diameter of the earth. Suppose two observers at opposite points on the earth were to measure the sun's position in the sky. The sun is some 93 million miles away, while earth's equatorial diameter is only about 8,000 miles, so the change in the sun's

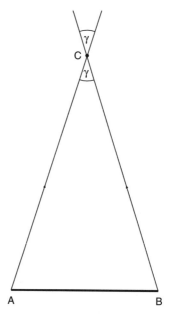

FIGURE 2 Stellar parallax.

position relative to the celestial sphere would be very small; but it could still be detected—*if* there were some reference stars visible in the background. Alas, during daytime the sun's glare washes out everything else in the sky, so a direct measurement of the sun's position relative to the fixed stars is impossible. This is where Edmond Halley stepped in: he proposed to use Venus instead of the sun—provided one was willing to wait for the rare occasion when the planet passes in front of the sun. Only four such occasions happened since Halley's time—in 1761, 1769, 1874, and 1882.

These four transits were the focus of a vast international effort aimed at obtaining an accurate value of the Earth-Sun distance—the *astronomical unit*. It became common among nineteenth-century astronomers to express this distance not in miles, but in parallax units. By astronomical convention, the *solar parallax* is *half* the angular shift in the sun's position, when using earth's diameter as baseline (the reason for taking half the angle is merely a mathematical convenience). Thus, if E_1 and E_2 (fig. 3) represent earth's diameter and S the sun, then the solar parallax is the angle α.[1] Determining this angle was the prime goal of the numerous expeditions sent to observe the four Venus transits; the most accepted value, obtained from the 1882 transit, was 8.857 arc seconds with a probable margin of error of ± 0.023 arc seconds. This translated into a solar distance of 92,300,000 miles, but the small margin of error still left an uncertainty of some 500,000 miles, to the chagrin of astronomers.[2] Fortunately, alternative methods were by then available, and these soon replaced the transit method.[3]

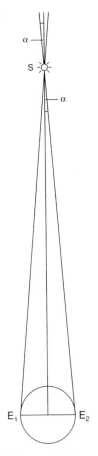

FIGURE 3 Solar parallax.

  ✷✷✷

EVEN earth's diameter, however, is insufficient for measuring distances of objects outside our solar system. For this task, a much longer baseline is necessary. Fortunately, such a baseline is available: the diameter of earth's orbit around the sun. The *stellar* (or *annual*) *parallax* is half the angular shift in a star's position when observed from two opposite points in earth's orbit—that is, at a half-year interval. Until the nineteenth century, all attempts to measure the parallax of a fixed star have failed: not a single star showed any apparent shift in position. The Greeks took this as proof of their geocentric worldview—an immovable Earth eternally fixed at the center of the universe. Copernicus interpreted the facts differently: to him, the absence of a discernible parallax meant that the fixed stars are at immense distances from us.

With the invention of the telescope in 1609, it became theoretically possible to look for the parallax of some nearby stars. But it was not until 1838 that Friedrich Wilhelm Bessel (1784–1846) succeeded in measuring the tiny parallax of the star 61 Cygni in the constellation Cygnus the Swan, known for its large proper motion. After eighteen months of intense observation, he announced that 61 Cygni showed a parallax of 0.314 arc seconds—about the angle a dime would subtend at a distance of 7 miles (by comparison, the moon's apparent diameter is about 1,800 arc seconds). This allowed him to fix the star's distance at 10.1 light years, or more than 60 trillion miles. To get some idea of how immense this distance is, imagine that we shrink the earth to the size of a cherrystone. The sun will then be represented by a basketball-size sphere 100 feet away, while 61 Cygni will be a slightly smaller sphere some 15,000 miles away. And 61 Cygni is only one of our closest stellar neighbors; most stars visible to the naked eye are hundreds or thousands of light years away.[4]

Just as with solar parallax, it has become common to express stellar distances in terms of parallax units. One *parsec* (short for "parallax second") is the distance to a star whose annual parallax is one arc second; or put differently, the distance at which the radius of earth's orbit subtends an angle of one arc second. One parsec is about 3.26 light years.

Because stellar parallaxes are so small, the method of parallax could, until recently, be applied only to a handful of nearby stars up to a distance of some 30 parsecs. But in 1989, the European Space Agency launched *Hipparcos*, a parallax-measuring satellite, and the field of astrometry (stellar position measurement) was at once revolutionized. *Hipparcos* is an acronym for High Precision Parallax Collecting Satellite, but the name was also chosen in honor of Hipparchus. In its four years of operation, it measured the parallax of some 118,000 stars to an unprecedented accuracy of 0.001 arc seconds. This tiny angle is about the size of a dime when viewed from 2,000 miles away; or to give a more dramatic example, the size an astronaut on the moon, if he could be seen by a terrestrial observer with unaided eyes.[5]

Notes and Sources

1. Alternatively, it is the angular width at which earth's semi-diameter would be seen from the sun.

2. The value 8.857 (± 0.023 arc seconds) is given by Simon Newcomb in his *Astronomical Constants* (1896), with other authors giving slightly different values. See Agnes M. Clerke, *A Popular*

History of Astronomy during the Nineteenth Century (1885; 3d ed. London: Adams and Charles Black, 1893), pp. 296–298; Simon Newcomb, *Popular Astronomy* (New York: Harper, 1880), pp. 183–184; and Charles A. Young, *A Text-Book of General Astronomy for Colleges and Scientific Schools* (1888; rev. ed. Boston: Ginn and Company, 1904), pp. 421–422.

3. Newcomb, in *Popular Astronomy*, devoted a full thirty pages to the various methods of determining the solar parallax (he also included an extensive list of papers published on the subject between 1854 and 1877). This attests to the importance that nineteenth-century astronomers attached to the problem. See also Clerke, *A Popular History of Astronomy*, part 2, chapter 6, and Young, *A Text-Book of General Astronomy*, chapter 17.

4. The parallax of 61 Cygni has since been refined to 0.294 arc seconds, resulting in a distance of 11.1 light years. The nearest known star system is Alpha Centauri, a triple star system whose closest component, Proxima Centauri (discovered in 1915), is at present 4.2 light years away. See the article, "Our Nearest Celestial Neighbors," by Joshua Roth and Roger W. Sinnott, *Sky & Telescope*, October 1996, pp. 32–34.

5. See the article, "Measuring the Universe: From Hipparchus to Hipparcos," by Catherine Turon, *Sky & Telescope*, July 1997, pp. 28–34.

5

The Dance of Two Planets

It was Homer who, 3,000 years ago, saluted Venus as
the most beautiful of stars. Who has not been struck
with her wonderful brilliance? Who can refrain, when
she shines so marvelously in the heavens, from greet-
ing her as the brightest of stars and asking what mys-
teries are hidden in that light?

Camille flammarion, *Dreams of an Astronomer*
(1923)

NO SCIENCE, in the eyes of the public, is held in higher es-
teem than astronomy. Not physics, which aims to unravel
the fundamental laws governing our universe; not biology,
striving to understand life on this planet and elsewhere; not
geology, oceanography, and meteorology, whose findings
help us preserve—and also exploit—our own planet, its
lands, waters, and atmosphere; not even mathematics,
without which a serious study of nature would be impossible. Of all the sciences, it is astronomy, whose direct impact
on our daily lives is minimal, that has always fired people's
imagination. Its predictions are being held by the public
with absolute faith. When a celestial event, calculated years
in advance, occurs at the exact appointed place and time,
the astronomer who made the prediction is showered with
accolades. But when a heavenly forecast fails to materialize,

as sometimes happens, the integrity of the entire profession is put in doubt.

Young Tycho Brahe, Kepler's great predecessor and mentor, was so impressed by the onset of a solar eclipse that he decided then and there to devote himself to astronomy. This happened in 1560, when he was just fourteen years old. The eclipse was only partial in his native Denmark— an interesting but by no means spectacular event—but what impressed Tycho was the predictability of it, "that men could know the motions of the stars so accurately that they were able a long time beforehand to predict their places and relative positions."[1]

When the eighth planet, Neptune, was discovered in 1846 at the exact location in the sky where French astronomer Urbain Leverrier had calculated it should be, the world was agog with excitement. But half a century later, a singular flop almost undid the glory: following three spectacular meteor showers in November of 1799, 1833, and 1866, astronomers confidently predicted that the show will repeat in November 1899. But when the promised spectacle failed to occur, embarrassed astronomers had to endure the public's ridicule; it was, in the words of American meteor scientist Charles P. Olivier, "the worst blow ever suffered by astronomy in the eyes of the general public."

On at least one occasion, the failure to predict an astronomical event had dire consequences. As told in the Chinese classic *Shu Ching*, a solar eclipse occurred over China during the reign of Chung K'ang. Hsi and Ho, the court officials in charge of heavenly events, having been too fond of wine, had neglected their duties and failed to announce this important event, a crime for which they were condemned to death. This eclipse has been identified as having occurred on October 22, 2137 B.C.

* * *

How, then, can astronomers predict the occurrence of a transit hundreds of years before it actually happens? And why are transits—and those of Venus in particular—so rare? To answer these questions, we must examine the way Venus and Earth move around the sun. The basic facts have been known for centuries: Venus's orbit around the sun lies inside Earth's orbit, and consequently Venus's "year"—the time it takes the planet to complete one revolution around the sun—is only 224.7 days long, compared with Earth's 365.256 days. These figures lead to a curious coincidence: thirteen Venus "years" are almost exactly equal to eight Earth years (13 × 224.7 = 2921.1 days, whereas 8 × 365.256 = 2922.05 days). This coincidence has important bearings on Venus's transit schedule.

Imagine that at some given moment, Earth and Venus are lined up with the sun, with Venus between Earth and the sun. This alignment is known as *inferior conjunction* (by contrast, during a *superior* conjunction, Venus is on the opposite side of its orbit as seen from Earth, so that the sun is between the two planets). Venus, moving faster in its orbit than Earth, will quickly gain over the latter, so that after one year, when Earth has completed one orbit around the sun, Venus will be well ahead of Earth. After a second year, Venus will be ahead by twice as much, and with each subsequent year the lead will increase proportionately. But in an orbit, "ahead" soon becomes "behind"; after a certain time interval, Venus will overtake Earth from behind, and the two planets will once again be lined up with the sun.

The interval between two consecutive inferior conjunctions is called a *synodic period* (from the Greek *synodos*, a meeting or assembly), and we now set out to find it.

Let us, for the sake of argument, replace Earth and Venus with the small and large hands, respectively, of an old-fashioned clock (I say "old-fashioned" to distinguish it from the modern digital clock). At twelve o'clock, the two hands are overlapping. When will this happen again? We know from experience that this should happen shortly after one o'clock, but exactly when? This problem often appears in beginning algebra books, and it always befuddles students. But once you have the right idea, it really becomes a simple problem. The "right idea" is to look at the speeds, or *rates*, at which the two hands move around the clock. The large hand completes a full rotation in one hour, so its rate is one rotation per hour. The small hand completes a full rotation in twelve hours, so its rate is one-twelfth of a rotation per hour. The large hand thus overtakes the small hand with a relative rate of $(1 - \frac{1}{12})$, or $\frac{11}{12}$, rotations per hour.

Now, for the two hands to meet again, the large hand must be ahead of the small hand by exactly one rotation. Let us denote by T the time it takes the large hand to gain a full rotation over the small hand. Recalling the basic equation of uniform motion, distance = rate × time, we are led to the equation

$$\tfrac{11}{12} T = 1,$$

where the 1 on the right side stands for one rotation. Dividing both sides by $\frac{11}{12}$ (which is the same as multiplying by $\frac{12}{11}$), we get the required solution,

$$T = \tfrac{12}{11} = 1\tfrac{1}{11} \text{ hours.}$$

Now, we usually measure parts of an hour in minutes and seconds, so let us change the fractional part $\frac{1}{11}$ into minutes and seconds. Since one hour equals sixty minutes, $\frac{1}{11}$ of an hour equals $\frac{60}{11}$ minutes, or $5\frac{5}{11}$ minutes. And since one minute equals sixty seconds, $\frac{5}{11}$ of a minute equals $60 \times \frac{5}{11}$ $= \frac{300}{11} = 27.2727\ldots$ seconds. The "synodic period" of our clock, then, is one hour, five minutes, twenty-seven seconds, and a fraction of a second.

We now return to Venus and Earth. Since Venus completes a full orbit in 224.7 days, its rate of motion is 1/224.7 rotations per day. Similarly, Earth's rate is 1/365.256 rotations per day. Venus thus overtakes Earth with a relative rate of (1/224.7 − 1/365.256) rotations per day. With a calculator, we can easily find the decimal value of this difference, about 0.0017126. Using this number in the equation distance = rate × time, and again denoting the required time interval by T, we get

$$0.0017126\,T = 1.$$

This equation is entirely similar to our "clock equation"; only the coefficient of T is different. To solve it, we divide both sides by 0.0017126, again using a calculator. The result is

$$T = 583.9169\ \text{days},$$

or very nearly 584 days, which are about nineteen months. This, then, is Venus's synodic period—the time between two consecutive inferior conjunctions. This period was well known to the ancients; records of it appear in Babylonian and Mayan inscriptions, and the Mayan calendar may have been based on it.[2]

We mentioned earlier that thirteen Venus "years" are

almost exactly equal to eight Earth years. From this it follows—one would have to do a bit of simple algebra to show this—that five synodic periods of Venus are almost exactly equal to eight Earth years. Indeed, 5 × 583.9169 = 2919.5845 days, whereas 8 × 365.256 = 2922.048 days. *Thus after every eight years, Sun, Venus, and Earth occupy nearly the same position in space—that is, relative to the fixed stars.* It is important not to confuse this period with Venus's synodic period, after which the three bodies only return to the same position *relative to one another.* Again our clock analogy may help: the two hands overlap every one hour, five minutes, and twenty-seven seconds, but they return to their original position in "space" (i.e., relative to the clock's outer rim) only after twelve hours.

***

THE question now arises: Why doesn't a transit occur after each synodic period, when Venus is in inferior conjunction? The same question is often asked in connection with eclipses: Why doesn't a solar eclipse occur every month, when the "new" moon stands between the earth and sun? The answer to both questions is the same: the moon's orbit around the earth, and Venus's orbit around the sun, are slightly tilted to Earth's own orbit. Let us think of Earth's orbit as a large hula hoop lying on the floor, with the sun at the center. The plane of this orbit is called the *ecliptic;* it serves as a ground plane relative to which the orbits of all other planets are reckoned.[3] Earth completes one revolution around the sun once a year, moving in its orbit from west to east; but as viewed *from the earth,* the ecliptic is an imaginary circle, drawn on the celestial sphere, along which the sun seems to move slowly eastward from day to day,

completing one revolution in one year. This imaginary circle is the intersection of Earth's orbital plane with the celestial sphere.[4]

Imagine now a second hula hoop, smaller than the first
and lying inside it, but slightly tilted toward it. Of course, if
the large hoop lies on the floor, we have to imagine the small
hoop as extending halfway above the floor, and halfway
below it (fig. 5.1). This inner hoop represents Venus's orbit.
Now, in space there is no "above" or "below," but we still
need some word to distinguish between the two halves. We
call that half of Venus's orbit lying "above" the ecliptic the
northern half, and the half lying "below" it the *southern
half.* The tilt of Venus's orbit to the ecliptic is very small—
about 3 degrees and 23 arc minutes—but it is enough to im-
pose limitations on the times at which a transit can occur.

We need to introduce one more concept before we can
come up with a transit schedule. We know from geometry
that two planes—unless they are parallel to each other in
space—intersect in a straight line. This can be readily seen
when opening a book: the facing pages intersect at the line
where the book opens up. The plane of Venus's orbit around

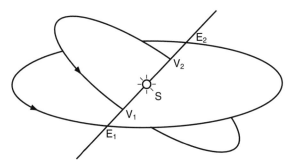

FIGURE 5.1 Venus's orbit relative to Earth's.

the sun, being tilted to the ecliptic, intersects it in a straight line, called the *line of nodes;* it passes through the center of the sun and divides Venus's orbit into a "northern" and a "southern" half (see again fig. 5.1). The line of nodes, over a long time period, is fixed in space: it keeps nearly the same direction relative to the fixed stars. As Venus moves around the sun, the planet will cross this line twice during each revolution: once going from south to north, and again going from north to south. These are the points where Venus's orbit crosses the ecliptic; they are called the *ascending node* and *descending node,* respectively.

Because of the slight tilt of Venus's orbit to the ecliptic, the planet, even when at inferior conjunction, is usually somewhat "above" or "below" the sun as seen from Earth, and a transit will not take place. *Clearly, a transit can occur only when Earth happens to cross the line of nodes at the same time as Venus.* At the present time—and for many centuries to come—Venus reaches its ascending node around December 8, and its descending node around June 7, so a transit can happen only around these dates. But for a transit *actually* to occur, Earth, too, must cross the line of nodes on these dates.

✳✳✳

IMAGINE now that these conditions have been fulfilled, and a transit has just occurred. What are the chances that a second transit will take place after eight years, when Sun, Venus, and Earth again occupy nearly the same position in space? Had the sun, as seen from Earth, been the size of a star—a point of light—the chances for a second transit (indeed, for *any* transit) would be infinitesimally small. Fortunately, the sun's disk has a sizable diameter—about 32 arc minutes, or roughly half a degree, as measured on the celestial sphere. Consequently, there is a little room for maneuvering, and a transit may still happen even if the three bodies are not exactly lined up in space. So there is at least a *possibility* that a second transit might occur eight years after the first. But will one actually happen?

To answer this question, we recall that five synodic periods of Venus are almost exactly equal to eight Earth years: $5 \times 583.9169 = 2919.5845$ days, whereas $8 \times 365.256 = 2922.048$ days. Indeed, had we rounded the two periods to the nearest full day, the match would have been exact ($5 \times 584 = 8 \times 365 = 2920$ days). But the small deviation from a full day causes Earth to lag behind Venus by about 2.46 days ($= 2922.048 - 2919.5845$) every fifth conjunction. This may not seem much, but during those 2.46 days, Venus has moved slightly higher or lower in its orbit relative to the ecliptic. Knowing Venus's orbital velocity and angle of inclination to the ecliptic, a fairly straightforward calculation shows that in 2.46 days, Venus, when near its ascending or descending node, is carried about 22 arc minutes higher or lower, respectively, in its orbit. Now the solar disk has a diameter of about 32 arc minutes, so its radius is about 16 arc minutes. Therefore, if during the earlier transit Venus crossed the solar disk exactly at its center, the next passage eight years later would carry it just outside the sun's disk, and a transit will not occur (fig. 5.2). But if the earlier tran-

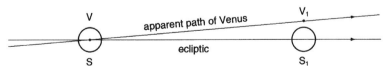

FIGURE 5.2 A single transit followed by a near miss eight years later.

sit was *not* central, but closer to the solar limb, then Venus has enough "clearance" to allow a second transit to happen (fig. 5.3). We conclude that *transits of Venus are likely to occur in pairs separated by eight years.* This is indeed the case at the present time.

During the two transits of a pair, an observer on the earth will see Venus cross the sun's disk along parallel lines, about 22 arc minutes apart (fig. 5.4). Can there be another transit after eight more years? Suppose the first transit occurred at the sun's lower (southern) limb when Venus was at its ascending node. Five synodic periods later—roughly eight years—Venus will be 22 arc minutes higher, which will still allow it to cross the solar disk. But five more synodic periods will shift the path another 22 arc minutes higher, resulting in a total shift of 44 arc minutes, so Venus will have just missed the sun by 12 arc minutes (= 44 − 32). Thus, *three transits separated by eight-year intervals cannot happen.*

So when *will* Venus transit the sun again? Under "ideal" circumstances—if Venus and Earth orbited the sun in perfectly circular orbits, and if no other planets were around to add their own gravitational pull on the two planets—another transit would occur after roughly $121\frac{1}{2}$ years. This is the theoretical time interval it takes Venus to go from its ascending node to its descending node and be again close enough to the center of the sun's disk to cause a transit. But

FIGURE 5.3 A transit pair separated by eight years. $N =$ Venus's ascending node.

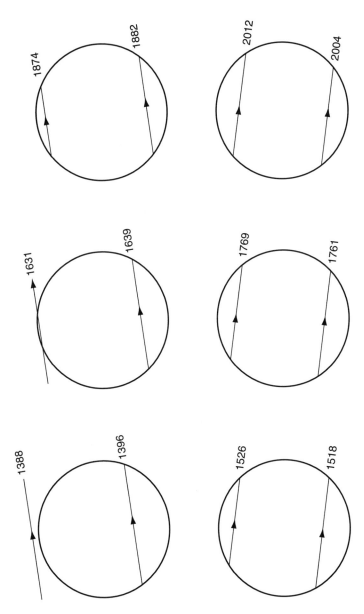

FIGURE 5.4 The eleven transits of Venus from 1396 to 2012.

at this point we must introduce a complication. Earth and Venus orbit the sun in elliptical, not circular orbits; and although the departure from a circle is very small (indeed, Venus's orbit is the most circle-like among the nine major planets), it nevertheless introduces a slight imbalance into the picture. For one, a planet does not move uniformly in its orbit; it moves faster at its closest point to the sun, and slower at its farthest point. And the sun's apparent diameter also varies: it is slightly larger when Earth is closest to the sun (which happens in January), and slightly smaller at its farthest point (in July). The combined effect of these factors is to move up the recurrence of a transit pair by sixteen years—that is, by ten synodic periods—at the ascending node (in December). We thus have the following transit schedule, beginning with a December transit:

A	A		D	D		A	A		D	D		A
8	$121\frac{1}{2}$	8	$105\frac{1}{2}$	8	$121\frac{1}{2}$	8	$105\frac{1}{2}$					

$\longleftarrow$ ——— 243 ——— $\longrightarrow$ $\longleftarrow$ ——— 243——— $\longrightarrow$

Here "A" denotes a transit at the ascending node, and "D" a transit at the descending node. We see that a complete transit cycle—the interval between the first transit of an "A" pair and the first transit of the following "A" pair—is 243 years.

***

THIS is the state of affairs at the present time. But it may happen that Venus just misses the sun when it "should" have passed in front of it. This happened in 1388, when a transit that should have occurred on November 26 (by the

Julian, or "old style" calendar then in use) failed to materialize (fig. 5.4). This non-event was followed eight years later by a single transit on November 23, 1396, the last in a long period of single November transits that began in the year A.D. 667; this period was characterized by the time intervals $121\frac{1}{2}$—8—$113\frac{1}{2}$. There are yet other possibilities. During the period from May 22, 427 B.C., to November 23, A.D. 424, *both* pairs in a transit cycle were replaced by single transits; this resulted in the cycle $121\frac{1}{2}$—$121\frac{1}{2}$. The present cycle of 8—$121\frac{1}{2}$—8—$105\frac{1}{2}$ began on December 7, 1631—the transit that Kepler had predicted and Gassendi had tried but failed to observe; it will end on June 14, 2984. We are thus fortunate to live in a double-transit period, giving most of us a chance to see two transits in a lifetime. Future generations will not be so favored: on December 18, 3089, a series of single December transits will begin and last until December 25, 3818; this series follows the cycle $129\frac{1}{2}$—8—$105\frac{1}{2}$. It is difficult indeed to find an orderly pattern in these cycles; the only constant feature seems to be the 243-year period of a complete transit cycle.[5]

✳✳✳

THERE are a few other transit-related features worth mentioning. A transit—either of Mercury or Venus—always begins on the eastern limb of the sun and ends on the western limb. This is in direct contrast to eclipses—either solar or lunar—which usually proceed from west to east. The reason for this is that Mercury, Venus, and Earth move around the sun in the same direction—from west to east, as seen from the sun. During a transit, Venus is at its inferior conjunction, so it is on the same side of the sun as Earth. And since Venus moves faster in its orbit than Earth, it will pass in front of the sun *from east to west* (fig. 5.5). This reversal of the "normal" movement of the planets is simply a result of the fact that we are watching the inner planets from our own moving platform; it is just another demonstration that all motion is relative.

Knowing the rates of motion of Venus and Earth and the distance between them at inferior conjunction, one can calculate the duration of a transit. A *central transit*—when Venus passes exactly through the center of the solar disk—lasts a little more than eight hours from first to last contact. The transit of November 23, A.D. 424, was nearly central; the next central transit—you might want to put the date on your calendar—will be on July 11, 5900. At the other extreme, on December 14, 2854, there will be a *grazing transit*, with Venus just touching the solar disk externally. In fact, if you happen to be in Antarctica on that day, you might witness a very small *partial transit*—Venus's black circle

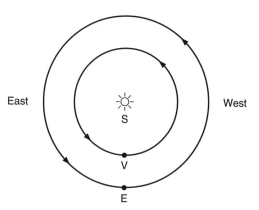

FIGURE 5.5 The relative positions of Venus and Earth during a transit.

briefly dipping into the solar disk before leaving it again; elsewhere on our globe, the transit will be a grazing one, or more likely, not visible at all.

We said earlier that at the present time, a transit of Venus can only happen within a few days on either side of December 8 or June 7—the approximate dates when Earth crosses Venus's line of nodes. But this line is not absolutely fixed in space. It performs a very slow *retrograde* motion (that is, from east to west) at the rate of about 20.5 arc seconds per year. At the same time, Earth's own axis of rotation, and with it the celestial equator, performs a retrograde motion at the rate of 50.3 arc seconds per year (this slow wobbling motion, known as the precession of the equinoxes, is caused by the gravitational pull of the sun and moon on Earth's equatorial bulge; it takes about 25,800 years to complete one cycle). Since the two motions are in the same direction, Venus's line of nodes will slowly move *forward* (from west to east), as seen from Earth, by about 29.8 (= 50.3 − 20.5) arc seconds per year. It will therefore take Earth a bit longer each year to cross Venus's line of nodes. Over a complete transit cycle of 243 years, the delay amounts to about two days; consequently, the dates on which a transit can occur are pushed back, on the average, by two days every complete transit cycle. This is illustrated in the following table, which gives the dates of three complete cycles:

Dec 7, 1631	Dec 9, 1874	Dec 11, 2117
Dec 4, 1639	Dec 6, 1882	Dec 8, 2125
Jun 6, 1761	Jun 8, 2004	Jun 11, 2247
Jun 3, 1769	Jun 6, 2012	Jun 9, 2255

There will come a day, sometime after the year 4000, when the December transits will actually occur in January.

But before that happens, there will be one more event to be marked on your calendar: on Christmas Day, December 25, 3818, Venus will pass almost centrally in front of the sun. This will be the longest transit since the year A.D. 424, lasting a total of eight hours and nine minutes, and it will be visible from nearly the entire Western Hemisphere. No doubt those around to watch it will give this Christmas Transit a special spiritual significance. Stay tuned!

Notes and Sources

1. J.L.E. Dreyer, *Tycho Brahe* (Edinburgh, 1890), as quoted by Arthur Koestler in *The Watershed: A Biography of Johannes Kepler* (New York: Anchor Books, 1960), p. 86.

2. On this subject, see Hugh Thurston, *Early Astronomy* (New York: Springer, 1994), pp. 199–201. See also Dick Teresi, *Lost Discoveries: The Ancient Roots of Modern Science—from the Babylonians to the Maya* (New York: Simon and Schuster, 2002), pp. 97–99 and 117. According to Teresi, the Mayans already knew that the morning and evening stars were one and the same object, the planet Venus.

3. The word *ecliptic* comes from *eclipse;* an eclipse—solar or lunar—can happen only when the moon, whose orbit is tilted to the ecliptic by about 5 degrees, happens to cross the latter.

4. It is along this circle that the twelve constellations of the zodiac are located, dividing the year into twelve convenient regions. The sun covers roughly one constellation each month.

5. The data on past and future transits is based on Jean Meeus, *Transits* (Richmond, Va.: Willmann-Bell, 1989), pp. 11–13 and 46–49. figure 5.4 is based on figure 5 in that book.

6

A Call for Action

> But Venus is very rarely seen within the sun's disk; and
> for a series of 120 years, is not to be seen there once;
> viz. from 1639, when Mr. Horrox was favoured with
> this agreeable sight, the first and only one since the cre-
> ation, down to 1761; at which time Venus will pass over
> the sun on May 26 [Old Style] in the morning.
>
> Edmond Halley, *Philosophical Transactions* (1716)

ON A COSMIC SCALE, a century is but a fleeting moment.
But here on earth, much can happen in a century. Between
Venus's transit of 1639 and its return in 1761, the world had
changed as never before. Europe saw the rise of powerful
monarchs: Frederick the Great in Prussia, Maria Theresa
in Austria, and Louis XIV, the Sun King, in France. The
peace of Westphalia ended the bloody Thirty Year War,
Oliver Cromwell defeated the Royalists in England, and
Peter the Great moved his capital to St. Petersburg, thereby
opening his country to the West. In China, the Manchu
Ch'ing dynasty rose to power, while in the New World,
Philadelphia was founded and New Amsterdam became
New York. And in science, the towering figure of Isaac
Newton (1642–1727) would dominate the scene for the
next 250 years: his invention of the calculus was the most
profound advance in mathematics since ancient times, and

his universal law of gravitation, in effect, unified physics and astronomy.

Against this backdrop rose one the most versatile scientists of all time: Edmond Halley. Born near London on November 8, 1656, to a wealthy father who was murdered in an obscure dispute, Halley showed an interest in science at a young age. When only twenty years old, he sailed to the island of St. Helena in the South Atlantic (where the exiled Napoleon Bonaparte, a century later, would spend his last years) to conduct the first systematic survey of the southern skies.[1] On his return home he was elected to the Royal Society, and soon befriended Newton. He played a pivotal role in convincing the reclusive Newton to publish his great work, the *Principia,* in which he expounded his universal law of gravitation. Halley agreed to fund the publication out of his own pocket.

Halley's work covered an enormous range of topics. He designed an early version of the bathysphere, a device that allowed divers to extend the time they could spend underwater; he charted earth's magnetic field with the goal of using it to improve the accuracy of maritime navigation; he was the first to plot the path of a solar eclipse on a terrestrial map; and in mathematics, he pioneered the use of statistical methods in the analysis of mortality figures, setting the stage for modern insurance practices.

But it was his prediction of the return of the comet of 1682 that made Halley's name immortal. Newton's theory of gravitation had fully explained the motion of the then six known planets; but comets—those vagabond heavenly visitors whose unpredictable apparitions spread terror and fear of impending disasters—were a different matter. The sight of the comet of 1682 left a lasting impression on Halley. With Newton's help, he examined the records of past comets and came to a startling conclusion: the comets of 1456, 1531, 1607, and 1682 were one and the same celestial object, whose elongated elliptical orbit around the sun brought it close to the earth every seventy-six years. He predicted that the same comet would return again in 1758— give or take a few months, due to the gravitational pull of the giant planets Jupiter and Saturn, which might perturb it from its normal orbit. Halley, who was already fifty when he announced his prediction, knew he would not live to see it come true, but he reminded young astronomers that "should it return again about the year 1758, candid posterity will not refuse to acknowledge that this was first discovered by an Englishman." Halley's boast was not vain: his was the first time anyone had incorporated these mysterious heavenly wanderers into the framework of a coherent scientific theory. The celebrated celestial visitor did return on time: it was first spotted on Christmas Day, 1758, and it reached its closest point to the sun four months later. Halley's comet, as it has been known since, would faithfully keep its appointments in 1835 and in 1910; and when it returned again in 1986, it was greeted by a fleet of spacecraft sent out to intercept the famous wanderer and take close-up photos of it. Halley would have been delighted.

Halley's scientific output did not diminish with age. In 1718, when he was sixty-two, he announced another major discovery: some of the brightest stars in the sky had changed their positions since Ptolemy had recorded them in his star catalog in the second century A.D. This was the first concrete evidence that the "fixed" stars were anything but fixed; it is only because they are so far away from us that we cannot detect their motion over a short time span.

In 1719 John Flamsteed, the first Astronomer Royal, died, and Halley was appointed to replace him. In that capacity he greatly improved the Royal Observatory at Greenwich and then devoted himself to a meticulous study of the moon's motion, a subject of great importance at the time because of its potential use for determining the longitude of a ship at sea. He continued his work there until his death in 1742 at the age of eighty-five.

✱✱✱

ON November 7, 1677, while he was stationed on the island of St. Helena, the young Halley observed a transit of Mercury; this was only the fourth transit of the innermost planet ever seen, and the first to be observed from beginning to end. It was at this event that he got the idea of using the occasion of a transit to determine the astronomical unit, the mean Earth-Sun distance. The concept was not entirely new: the Scottish mathematician James Gregory (1638–1675) had already proposed it in 1663. But while Gregory's was just a concept, expressed in general terms, Halley came up with a detailed plan that could actually be carried out. It would germinate in his mind for the next forty years. In 1716, when he was sixty years old, he submitted his proposal to the Royal Society. It opened with words reminiscent of Horrocks's admonition to astronomers a century earlier:

> There are many things exceedingly paradoxical, and that seems quite incredible to the illiterate, which yet, by means of mathematical principles, may be easily solved. Scarce any problem will appear more difficult than that of determining the distance from the sun to the earth, very near the truth; but even this, when we are made ac-

quainted with some exact observations, taken at places fixed upon and chosen beforehand, will, without much labour, be effected. And this is what I am desirous to lay before this illustrious Society (which I foretell will continue for ages), that I may explain beforehand to young astronomers, who may perhaps live to observe these things, a method by which the immense distance of the sun may be truly obtained to within a five-hundredth part of what it really is.[2]

Halley then went into the technical details of his proposal. We need not follow these details here;[3] it will suffice to give the main idea of his plan, which was to use the passage of Venus in front of the sun as a means to measure Venus's parallax, from which its distance from the earth could be computed. And once this distance was known, all other distances in the solar system—and in particular, the value of the astronomical unit—would easily follow, using Kepler's third law.

We saw earlier how the distance to a faraway object can be found from its parallax—the apparent shift in the position of the object relative to a remote background, when the observer changes his own position, or when viewed simultaneously by two observers from different places. The closer the object, the larger its parallax, and the greater the accuracy in finding its distance. Of all nine planets, the one that comes nearest to us is Venus; at its closest approach, it is only about 25 million miles away.[4] This should have made Venus the ideal candidate for parallax measurements. But Venus's proximity also causes it to shine with dazzling brilliance; it is, in fact, the third brightest object in the sky after the sun and moon, and the only other object that can be

seen with the unaided eye in broad daylight (provided one
knows exactly where to look). This makes it extremely difficult to measure Venus's position relative to the fixed stars
at night—the background stars would simply be washed
away in its glare. But during a transit, the planet's outline is
etched against the solar disk with razor-sharp clarity. Halley's idea was to use the sun as a remote billboard on which
Venus's black silhouette could be marked with pinpoint
precision.

His plan was to have the transit observed from a large
number of stations, widely separated in longitude and latitude. From each station, the planet would appear to cross
the sun's disk along a slightly different path (fig. 6.1). If each
observer could track Venus's path as seen from his station,
a comparison of the different paths would give the angular
separation between them; and this, combined with the
known location of the observers, would result in a series of

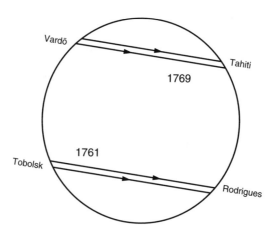

FIGURE 6.1 The transits of 1761 and 1769, each
observed from two different stations.

parallax values, which could then be averaged to get a single parallax.

It sounded simple, but there was a caveat: to track Venus's path, one would have to record the exact angular positions of the points of entry and exit relative to the sun's disk. In Halley's time, techniques for measuring angular positions were still primitive and unreliable, and this would have compromised the goal of the entire endeavor. Halley therefore proposed an indirect approach. Instead of tracking the actual paths, each observer would record the exact moment when Venus's disk has fully entered the sun (the moment of ingress), and again the moment when it is about to leave the sun (egress). The elapsed time between the two moments gives the duration of the transit for each location. Knowing the duration and Venus's rate of motion, one could find the length of both paths and the angular separation between them.

The success of Halley's method depended on two factors: an exact knowledge of the geographical position, in terms of longitude and latitude, of each station, and a precise timing of the moments of ingress and egress. The first factor presented relatively few difficulties and could be achieved with standard navigational equipment, either before or after the transit; the timing, however, was an all-or-nothing endeavor: it had to be achieved in "real time," while the transit was going on. Halley estimated that if the moments of ingress and egress could be timed to the nearest second, the astronomical unit could be found with an accuracy of one part in five hundred.

But why not use Mercury for this purpose, whose transits are much more frequent than those of Venus—on the average, thirteen every century? The reason, as Halley pointed out, is that Mercury is too close to the sun to be an effective target for parallax measurements. Stretch your arm with your thumb raised, and watch it first with your right eye, then with your left; if you hold your thumb close to the wall, the apparent change in position will hardly be noticeable. Only Venus, almost three-fourths the distance from the sun as Earth, would hold any chance of success. But this advantage came at a price: one had to wait a long time until a transit occurred—either eight years, if one was lucky to watch the first of a transit pair, or else more than a century.

Having outlined the general features of his plan, Halley then recommended several locations as potential observing sites for the upcoming transit of 1761; these were chosen with the goal of achieving the greatest difference in the time it would take Venus to cross the sun's face—up to seventeen minutes for locations at the North and South Poles, compared to a total duration of over five hours for the entire transit. Of course, the various stations also had to have a reasonable chance for clear skies on the crucial day. It would, in short, have to be an international endeavor, and Halley called for other nations to join forces:

> If the French should be disposed to take any pains herein, an observer may station himself conveniently enough at Pondicherry, on the west shore of the Bay of Bengal, where the altitude of the [celestial] pole is twelve degrees. As to the Dutch, their celebrated mart at Batavia will afford them a place of observation fit enough for the purpose, provided they also have but a disposition to assist in advancing the knowledge of the heavens.

Reading these words, one can hardly fail to notice Halley's low opinion of his European neighbors—especially France, England's arch rival.

He concluded his memoir to the Royal Society with these prophetic words:

> I recommend it therefore again and again to those curious astronomers who, when I am dead, will have an opportunity of observing these things, that they would remember this my admonition, and diligently apply themselves with all their might in making this observation, and I earnestly wish them all imaginable success: in the first place, that they may not by the unseasonable obscurity of a cloudy sky be deprived of this most desirable sight, and then, that having ascertained with more exactness the magnitudes of the planetary orbits, it may redound to their immortal fame and glory.

When the young and unknown Horrocks issued his admonition in 1639, only one person, William Crabtree, had heeded his call; but now the call came from Edmond Halley, Astronomer Royal and world renowned for the comet whose return he had predicted. This time astronomers from all over the globe, supported by their governments, would be ready at their telescopes. It was to be the first truly international scientific endeavor ever attempted.

Notes and Sources

1. For an account of Halley's activities in St. Helena, see Joseph Ashbrook, *The Astronomical Scrapbook: Skywatchers, Pioneers, and Seekers in Astronomy* (Cambridge, Mass.: Sky Publishing Company, and Cambridge, U.K.: Cambridge University Press, 1984), chapter 42.

2. The quotations in this chapter are taken from Richard A. Proctor, *Transits of Venus: A Popular Account of Past and Coming Transits* (1874; 4th ed. London: Longmans, Green, and Co., 1882), chapter 2. See also Harlow Shapley and Helen E. Howarth, *A Source Book in Astronomy* (New York: McGraw-Hill, 1929), pp. 96–100.

3. See Appendix 1 for a more detailed presentation of Halley's method.

4. When at its closest, Venus's crescent outline can easily be seen with small binoculars; some observers even claim to have seen the crescent with unaided eyes.

7

Venus Returns

> This sight [of Venus on the sun], which is by far the no-
> blest astronomy can afford, is denied to mortals for a
> whole century, by the strict laws of motion. . . . By this
> observation alone, the distance to the sun from the
> earth might be determined.
>
> Edmond Halley, *Philosophical Transactions* (1691)

EDMOND HALLEY died on January 14, 1742, almost exactly
one hundred years after Newton's birth; he was eighty-five
years old, one year older than Newton when he died in
1727. Their combined life spanned the most productive
century in the history of science to date. The two were op-
posites in many ways: Newton was reclusive, humorless,
suspicious, and vicious in pursuing his rivals; Halley was
gregarious, sociable, and loved by his many colleagues.
Newton came from a humble, poor rural family whose father died before he was born; Halley was born to a wealthy
London family whose head was a successful landlord and
businessman (in 1684 his father was murdered while taking
a walk not far from their home; the cause or perpetrator
have never been discovered). Newton never left his native
country; Halley was a world traveler eager to explore new
places. Yet for all their differences, they were trusted
friends, and we owe it to Halley that Newton's great mas-

terpiece, the *Principia*, saw the light of day. Newton had erected a grand theory of the universe, in which everything was in a perpetual state of motion brought about by the force of gravity; Halley took his master's theory and successfully used it to predict future heavenly events. Yet he did not live to see his predictions come true: his comet returned in 1758, sixteen years after his death; and it would be three more years before Venus appeared again on the face of the sun, its first transit since 1639.

Meanwhile, preparations went ahead to observe the upcoming transit with the full attention Halley had called for. The first item of business was to review the details of Halley's plan. Surprisingly, this was undertaken not by the Royal Society, but by a Frenchman, Joseph-Nicolas Delisle (1688–1768). Delisle was the ninth child of Claude Delisle, a historian-geographer, and was turned to astronomy by watching a solar eclipse in 1706. He set up his own observatory in the Luxembourg Palace in Paris, only to be kicked out by the Duc d'Orleans, who decided to house his daughter, the Duchess de Berry, there; only after her death twelve years later was Delisle allowed to return to his old headquarters. But he still did not have an official employment, and to earn a living he gave mathematics classes and drew astrological forecasts—not an uncommon way for an astronomer to make ends meet at the time (Kepler, as we remember, did the same). Finally, in 1718, he was appointed as chair of mathematics at the Collège Royal in Paris. Three years later, Peter the Great invited him to found an observatory in Russia; he planned to stay there four years but stayed twenty-two, during which he trained many young astronomers and widened his circle of professional connections. He was thus the ideal person for the task now before him.

Reviewing Halley's plan for the upcoming transit, Delisle found that Venus's path across the sun would actually take place some six arc minutes lower than Halley had predicted—about one-fifth of the sun's apparent diameter (fig. 7.1). This may not seem like much, but it meant that the duration of the transit would be considerably shorter than Halley had predicted—by about an hour and twenty minutes. This was a serious setback to existing plans, for it eliminated several of the locations Halley had recommended. But Delisle also showed that it was not really necessary to record the duration of the entire transit from beginning to end; it would suffice to record the moments of internal contact at ingress or egress at each station. Many years before, in 1724, he went to London to discuss the issue with Halley and the aging Newton. Delisle had origi-

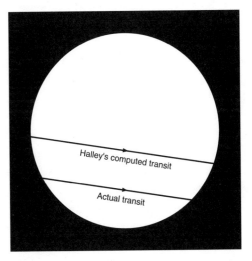

FIGURE 7.1 Halley's predicted versus actual path of the 1761 transit.

nally advocated using Mercury, rather than Venus, for determining the astronomical unit, because of the greater frequency of its transits; but his discussions with Halley, as well as the disappointing results from the 1753 Mercury transit, convinced him of the superiority of Venus transits. Now, in August 1760, less than a year before the event, he announced his method of observation, known since as "Delisle's method."

Delisle now became the unofficial coordinator and chief driving force behind the efforts to observe the upcoming transit. He published a large map that showed the regions from which the transit would be visible, either in its entirety or only partially. This map (a similar one of which is shown in fig. 7.2) reached the Royal Society in June 1760, barely a year before the scheduled event, and finally the British sprang into action.

The Society decided to send "proper persons to proper places" to observe the transit; but as the crucial date drew near, disagreements over funding and staffing the various expeditions threatened to undermine the whole project. The powerful East India Company at first agreed to fund an expedition to the island of St. Helena, sight of the 1677 Mercury transit that Halley had observed; but it refused to take part in a second expedition planned for Sumatra, fearing that the Seven Year War between England and France might put the participants in danger. James Bradley (1693–1762), who succeeded Halley as Astronomer Royal, then made up a list of the required equipment, which included navigational instruments, telescopes, pendulum clocks, and the personnel to operate them. By now the costs had gone beyond what the Society was willing to spend. In despera-

FIGURE 7.2 A mappemonde of the 1761 transit.

tion, a petition was sent to the admiralty requesting that it provide the means of transportation. Through intervention of various power brokers, the matter was finally settled, and the Society was issued a Royal Warrant for £800 to fund two expeditions. The final arrangements were left to Nevil Maskelyne (1732–1811), who would later become the fifth Astronomer Royal.

At last the two expeditions were ready. The first, headed by Maskelyne himself, sailed to St. Helena. The aristocratic Maskelyne refused to avail himself of the ship provided by the admiralty, preferring instead to travel on a private vessel. He was absent from England for sixteen months, during which he made some useful observations of ocean currents and tides; but as to his main mission, let me quote astronomer Simon Newcomb:

> Cloudy weather prevented his observations of the transit, but this did not prevent his expedition from leaving for posterity an interesting statement of the necessaries of an astronomer of that time. His itemized account of personal expenses was as follows:
>
> | | | | |
> |---|---|---|---|
> | One year's board at St. Helena: | £109 | 10s. | 0d. |
> | Liquors at 5s. per day: | 91 | 5 | 0 |
> | Washing at 9d. per day: | 13 | 13 | 9 |
> | Other expenses: | 27 | 7 | 6 |
> | Liquors on board ship for six months: | 50 | 0 | 0 |
> | | £291 | 16s. | 3d. |
>
> Seven hundred dollars was the total cost of liquors during the eighteen months of his absence. . . . He was subsequently Astronomer Royal of England for nearly half a century, but his published observations give no indication of the cost of the drinks necessary to their production.[1]

The second British expedition, headed for Sumatra (in modern Indonesia), was under the command of Bradley's assistant, the astronomer Charles Mason (1730–1787), who in turn was assisted by a young surveyor, Jeremiah Dixon (fl. 1763). The two would later become famous for marking the Mason-Dixon Line between Pennsylvania and Maryland, but right now they had trouble even getting on their way: their ship, the HMS *Sea Horse,* had hardly left Portsmouth when a French frigate intercepted them. In the ensuing gunfire, eleven of their sailors were killed and the ship was heavily damaged, forcing them to return to port. Badly

PLATE 1. Joseph-Nicolas Delisle.

PLATE 2. Jean Chappe d'Auteroche.

PLATE 3. Father Maximilian Hell.

PLATE 4. Alexandre-Gui Pingré.

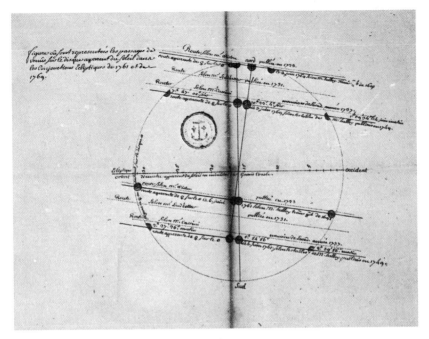

PLATE 5. The transits of Venus, 1761 and 1769.

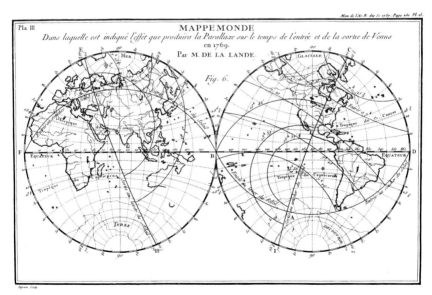

PLATE 6. Mappemonde for the transit of 1769.

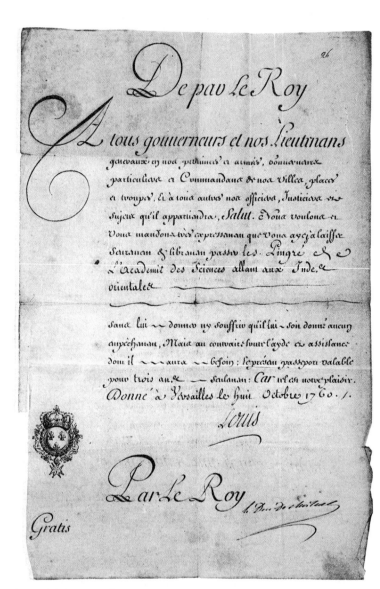

PLATE 7. Pingré's *laissez-passer*, 1761.

PLATE 8. Mural showing Crabtree observing the transit of Venus on December 4, 1639.

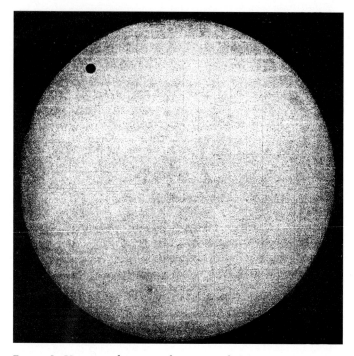

PLATE 9. Venus on the sun at the transit of 1874.

shaken, Mason informed the Royal Society that "we will not proceed thither, let the Consequence be what it will." The Society replied in no uncertain terms, reprimanding them for their change of heart and appealing to their sense of national honor: "Your refusal to proceed upon this Voyage, after having so publickly and notoriously ingaged in it . . . would be a reproach to the Nation in general, to the Royal Society in particular, and more Especially and fatally to yourselves"[2] The harsh words, backed up by a threat of legal action, had their desired effect: the group set sail again in February 1761. On April 27 they arrived at the Cape of Good Hope. Mason then engaged in a second act of disobedience: he decided to observe the transit from the Cape, notwithstanding his orders to go to Sumatra. He had a good reason: the town of his original destination had just been taken by the French. And so it was: they stationed themselves at the Cape and watched the entire transit in good weather; it was the only successful observation in the Southern Hemisphere.

***

THE French organized not two but four expeditions: one to Siberia, another to Vienna, and two to the southern seas. The first of the southern groups, headed by a French astronomer, Alexandre-Gui Pingré (1711–1796), was to sail to the island of Rodriguez, some 800 miles east of Madagascar in the Indian Ocean. The French and British were then at the height of their Seven Year War, but in a rare show of international goodwill the admiralty in London issued Pingré with a letter of safe passage, requesting all captains of His Majesty's ships to ensure that "Monsieur Pingré should not meet with any Interruption either in his passage to or from that Island. . . . You are hereby most strictly required and directed not to molest his person or Effects upon any Account, but to suffer him to proceed without delay or Interruption."[3]

Pingré reached his destination in May 1761, set up his equipment, and waited. On the crucial day, June 6, 1761, it rained all morning, but later there was a partial clearing, and he was able to observe the middle of the transit. But then the clouds set in again, and he failed to see Venus's exit from the sun. He thus missed the two most critical moments of the event, upon which the success of the entire endeavor depended. As if to add insult to injury, the few observations he did make were hampered by high winds which affected his delicate clocks.

On the way home his ship was molested by a British warship, whose captain ignored Pingré's letter of free passage. Repeatedly trying to evade the British, he was forced to disembark in Lisbon and complete the last leg of his journey by land over the Pyrenees. Finally back in Paris, he devoted himself to a project he had been commissioned to do back in 1756—design a sundial for the corn market in Paris which would show not only the hour, but also the sun's entrance onto each of the twelve constellations of the zodiac. Pingré spent his remaining years computing tables of past and future eclipses from 1000 B.C. to A.D. 1900, with the goal of assisting scholars in dating historic events. In 1783–1784 he published a four-volume work on comets, their nature, history, and orbits. His final work, a book on astronomical records of value to historians, was thirty years in the making; when it finally went to the printer, the project ran out of money, and many of the already printed pages were sold as scrap paper. Fortunately these were later recovered, and the complete work was published in 1901, more than a century after Pingré's death at the age of eighty-five. Most of his voluminous writings—including historical essays, poetry, fiction, and notes on practically every subject he dealt with—remain unpublished.[4]

***

PINGRÉ's experiences were but a slight inconvenience compared to the misfortunes of the second French expedition to the south. Headed by a gentleman with the long name Guillaume-Joseph-Hyacinthe-Jean-Baptise Gentil de la Galaisière—commonly known as Le Gentil (1725–1792)—its destination was the town of Pondicherry in India, some 100 miles south of Madras on the west coast of the Bay of Bengal. Le Gentil left France on March 26, 1760, more than a year in advance of the transit. On July 10 his ship stopped at the island of Mauritius—then known as Isle of France—east of Madagascar in the Indian Ocean. There he received bad news: the French plantation near Pondicherry, where he had intended to set up his instruments, had just been taken by the British. Undaunted, he changed his plans and resolved to reach the island of Rodriguez, further to the east, where Pingré was headed. As he was setting sail to his new destination, he learned that a French frigate was about to leave Mauritius for the coast of Coromandel in India, so he changed his plans again. But the frigate, *La Sylphide,* did not leave port until mid-March 1761—less than three months before the transit. Time was now running out. To make things worse, en route they experienced persistent calms, and Le Gentil was on the verge of despair. Finally, on May 24, their ship reached India. But then the captain received news that Pondicherry itself had just been taken by the English. So the captain turned his boat around and sailed straight back to Mauritius. For the remaining of the story, I quote from Proctor:

> The 6th of June arrived! The frigate was in 87° East longitude (from Paris) and 5° 45′ South latitude. The sky was clear, the sun splendid! The unfortunate Le Gentil, unwilling to be altogether idle, observed the transit on board the ship, taking all possible care. He noted the time of ingress and egress; but with what degree of approximation were those times obtained, even admitting that those he noted coincided exactly with the instant of the contacts? The voyage of the French academician ended thus in failure. . . . To have traversed so large a portion of the globe, to have endured all the weariness, all the privations, all the perils of a long sea-voyage, and to effect nothing! This was enough to have disgusted anyone.[5]

One would imagine him heading home, totally exhausted and demoralized. But not Le Gentil! He resolved to make the best out of a bad situation. Having an interest in the wildlife of the island, he decided to stay there, use the time to do some exploration of his own, and wait it out until the next transit eight years hence. We shall soon see what befell him there.

★★★

COMPARED to the adventures of the two French expeditions to the south, those of their northern colleagues were almost benign. It took the Abbé Jean-Baptiste Chappe d'Auteroche (1728–1769), head of the French expedition to Siberia, some five months to reach his destination near Tobolsk, east of the Ural Mountains. He spent the night be-

fore the transit at his station on a mountaintop outside the
town, "worrying about every distant cloud or smoke curl
from a fire."[6] But luck was with him, and he watched the
entire transit—5 hours, 48 minutes and 32.5 seconds in
all—in brilliant sunshine.

César-François Cassini de Thury (1714–1784), director
of the Paris Observatory and the third generation of a remarkable French family of astronomers, had the easiest
destination of all: Vienna, Austria. There, from the com-
fortable quarters of the Vienna Observatory, he watched
the transit in the company of Archduke Joseph of Austria
and the observatory's director, Father Maximilian Hell.
Hell would later be the center of a bitter controversy in
which he was suspected of falsifying the results of his observations so as to make them agree with those of the other
expeditions; in the end he was vindicated, but not before his
reputation was badly damaged (see page 126).

Some observers didn't even have to leave home. Stationed right in his office at the Royal Observatory in Greenwich, just outside London, was the Reverend Nathaniel
Bliss (1700–1764), who briefly succeeded Bradley as fourth
Astronomer Royal before his untimely death. Let me again
quote from Proctor:

> Early in the morning, when every astronomer was pre-
> paring for observing the transit, it unluckily happened
> that, both at London and Greenwich, the sky was so over-
> cast as to render it doubtful whether any part of the tran-
> sit would be seen; it was 38 minutes 21 seconds past 7
> o'clock, apparent time, at Greenwich when the Rev. Mr.
> Bliss, our Astronomer Royal, first saw Venus on the
> sun. . . . From that time to the beginning of egress the

Doctor made several observations, both of the difference of right ascension and declination [the celestial equivalents of longitude and latitude] of the centres of the sun and Venus, and at last found the beginning of egress to be at 8 hours 19 minutes 0 seconds apparent time. By the means of three good observations the diameter of Venus on the sun was 58 seconds of a degree [about one-thirtieth of the sun's apparent diameter].[7]

While the Astronomer Royal was intently observing Venus from Greenwich, one Mr. Short, at Savile House in London, was busy entertaining some dignitaries to the show. He watched the transit

in the presence of His Royal Highness the Duke of York, accompanied by Their Royal Highnesses Prince William, Prince Henry, and Prince Frederick. We are not told whether the Duke of York actually honoured Venus by directing His Royal gaze upon her during her transit, or whether Their Other Royal Highnesses made any observations; but as Venus was under observation for about $3\frac{1}{2}$ hours, we may suppose that these exalted persons did not lose the opportunity of witnessing a phenomenon so seldom seen. Venus, all unconscious of the honour, moved onward to egress, contact occurring at 8 h. 18 m. $15\frac{1}{2}$ s. apparent Greenwich time, or $8\frac{1}{2}$ s. sooner than at Greenwich.

    ✴✴✴

ONE observer made a historic discovery during the transit, although it took the world a hundred and fifty years to give him the credit he was due. Mikhail Vasilievitch Lomonosov (1711–1765) was born near the arctic port of Archangel in Russia, the son of a fish trader. Wishing to get a better education than his native place could offer, he made his way to Moscow and was admitted to school by claiming he was the son of a nobleman. At first aspiring to become a linguist, he eventually turned to science, completing his education in Germany and graduating in chemistry. Lomonosov worked in many fields, including physics, chemistry, geography, and astronomy. He was an early advocate of the theory that heat was a form of mechanical energy, that matter consisted of individual atoms, and that light propagated as waves—years before these ideas became part of mainstream physics. And in what must have been a harrowing experience, he and a friend tried to repeat Benjamin Franklin's famous kite experiment, designed to demonstrate that lightning is an electric discharge that could be conducted to the ground; his friend was killed, but Lomonosov escaped unharmed.

Lomonosov was soon elected to the Russian Academy of Science, and his reputation inside Russia grew. In 1755, together with the famous Swiss mathematician Leonhard Euler, he founded the University of Moscow. His early interest in linguistics never left him, though, and he proved himself a skillful poet and literary man who reformed Russian grammar and wrote the first comprehensive history of his country. In 1739 he wrote an ode commemorating the Russian victory over Turkey in the battle of Khotin. All this earned him the respect of his countrymen: his birthplace was named after him in 1948, as was a crater on the far side of the moon, first photographed by the Soviet spacecraft *Luna 3* in 1959. But in the West his name, even today, is hardly known; most histories of astronomy ignore him, and even the discovery that concerns us here has been attributed to others.[8]

It was during the 1761 transit that Lomonosov, observing from his home in St. Petersburg, saw a faint, luminous ring around Venus's black image just as it entered the sun's face (fig. 7.3); the sight was repeated at the moment of exit. He immediately interpreted this as due to an atmosphere around Venus, and he predicted that it might even be thicker than Earth's. Lomonosov reported his finding in a paper which, like most of his written work, was only published many years after his death in 1765. But it was not until 1910, one hundred and fifty years after the transit, that his paper appeared in German translation and became known in the West. Up until then the discovery of Venus's atmosphere had been credited to William Herschel, dis-

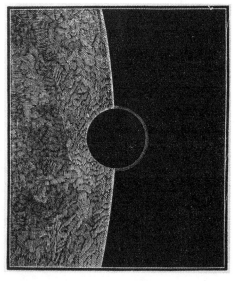

FIGURE 7.3 A halo around Venus just before ingress.

coverer of the planet Uranus; Lomonosov's paper makes it clear that he preceded Herschel by some thirty years.

✱✱✱

THE transit of 1761 was now history. It was observed from some seventy stations spread around the globe, the first large-scale international scientific endeavor ever attempted.[9] It was time to compare the many observations, analyze them mathematically, and, it was hoped, determine the solar distance with an unprecedented accuracy.

Notes and Sources

1. Simon Newcomb, *Reminiscences of an Astronomer* (Boston and New York: Houghton, Mifflin and Co., 1903), p. 153.

2. Willy Ley, *Watchers of the Skies: An Informal History of Astronomy from Babylon to the Space Age* (New York: Viking Press, 1966), p. 172.

3. Ibid., pp. 173–174.

4. More on Pingré's life can be found in Joseph Ashbrook, *The Astronomical Scrapbook: Skywatchers, Pioneers, and Seekers in Astronomy* (Cambridge, Mass.: Sky Publishing Corporation, and Cambridge, U.K.: Cambridge University Press, 1984), chapter 82.

5. Richard A. Proctor, *Transits of Venus: A Popular Account of Past and Coming Transits* (1874; 4th ed. London: Longmans, Green, and Co., 1882), pp. 55–56.

6. Ley, *Watchers of the Skies*, p. 175.

7. Proctor, *Transits of Venus*, pp. 51–52.

8. His life is told in Boris N. Menshutkin, *Russia's Lomonosov* (Princeton, N.J.: Princeton University Press, 1952).

9. A detailed list of these stations appears in Harry Woolf, *The Transits of Venus: A Study of Eighteenth-Century Science* (Princeton, N.J.: Princeton University Press, 1959), pp. 135–140.

8

A Second Chance

> Just before Venus reached the sun, the clouds gath-
> ered, and a storm burst upon the place. It lasted until
> the transit was over, and then cleared away again, as
> if with the express object of showing the unfortunate
> astronomer how helpless he was in the hands of the
> elements.
>
> Simon Newcomb, *Reminiscences of an Astronomer*
> (1903)

ALL SAID, the results of the 1761 transit observations fell short of the high expectations astronomers had put in them. Halley had hoped that the transit would allow astronomers to determine the mean Earth-Sun distance with an accuracy of one part in five hundred. The actual distance, as deduced from the various observations, ranged from 96,162,840 miles to 77,846,110 miles—far short of Halley's goal. One reason was the weather: at many locations it just did not cooperate. The irony of it could not be greater: the science that can predict the exact moment of an eclipse or transit a hundred years in advance is totally helpless in predicting the local weather even hours before the onset of the event. I had personally experienced this in 1991. Our group of amateur astronomers flew to Hawaii to watch the Great Solar Eclipse of July 11, one of the longest in this century.

Our destination was the Kona coast of the Big Island, chosen because of its superb weather: the month of July almost always sees clear skies, with favorable winds prevalining, on the average, 95 percent of the time. But the word "average" can be misleading, and on that day it was: to our great frustration, eclipse day dawned completely overcast, and except for a general darkening of the landscape around us, we saw nothing of the spectacle. To add insult to injury, on the day before the eclipse, *and* the day after, not a single cloud appeared in the sky! The unwelcome overcast on the crucial day was blamed on what local weather forecasters refer to as "TUTT"—a Tropical Upper Tropospheric Trough—a rather rare weather disturbance for the normally tranquil month of July; but that knowledge did little to ease our bitter disappointment. I tried to be jovial about the whole thing, telling my colleagues they should ask for a refund from our travel agent; some were in such a bad mood, they took me seriously.

There was also the problem of determining the exact geographical location of the various stations—a key factor in implementing Halley's method. Today, thanks to a network of navigation satellites, the position of any point on earth can be determined with an accuracy of a few yards; but in the eighteenth century a mariner would be glad to be able to fix his position to the nearest fifty miles. It is relatively easy to determine one's *latitude*—the distance north or south from the equator; one only has to measure the sun's altitude above the horizon at local noon, or (in the Northern Hemisphere) the altitude of the Polar Star at night. But *longitude*—the distance east or west from some standard meridian—was a different matter.[1] The longitude problem was the most pressing navigational challenge that had faced mariners since they began to sail the high seas: countless lives were lost due to ships failing to reach their destination.[2]

But if bad weather and navigational errors were factors that one could expect, the observers encountered something else for which they were not ready at all. It had been expected that at the moment when Venus's circular image completely entered the sun's disk, there would be a sudden, instantaneous separation of the planet's trailing edge from the sun's limb. This was the most crucial moment of the entire transit; its exact timing was the very goal of the entire elaborate observation program, and all eyes, telescopes, and clocks were waiting for this instant to happen. But instead of a sudden separation, Venus's trailing edge seemed to linger on for a while, as if it hesitated before taking the final plunge. A kind of ligament was forming between Venus and the sun's limb, like a drop of water just before separating from a faucet (fig. 8.1). The same phenomenon happened again just before egress, when Venus's leading edge was about to leave the sun. This *black drop effect,* as it became known, was seen by all those who observed the beginning or end of the transit. They used various words to describe it:

> The Rev. Hirst, watching from Madras, India: "At the total immersion the planet, instead of appearing truly circular, resembled more the form of a bergamont pear . . . yet the preceding limb of Venus was extremely well defined. . . . [Just before egress] the planet was as black as ink, and the body truly circular . . . yet it was no sooner in contact with the sun's preceding limb, than it assumed the same figure as before."

> Dr. John Bevis of the Royal Society, observing from Kew near London: "The planet seemed quite entered upon

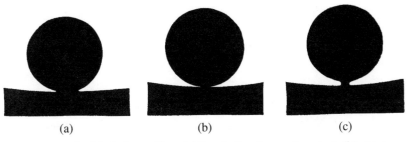

FIGURE 8.1 The black drop effect. (a) Just before internal contact; (b) at internal contact; (c) shortly after internal contact.

her disk, her upper limb being tangential to that of the sun; but instead of a thread of light, which he [Bevis] was expected immediately to appear between them, he perceived Venus to be still conjoined to the sun's limb by a slender tail, nothing near so dark as her disk, and shaped like the neck of a Florence flask. The said tail vanished at once; and for a few seconds after, the limb of Venus, to which it had been joined, appeared more prominent than her lower part, somewhat like the lesser end of an egg, but soon resumed its rotundity."

Mr. Pigott (location unknown): "Venus, before she separated from the sun, was considerably stretched out towards his [the sun's] limb, which gave the planet nearly the form of a pear; and even after the separation of the limbs, Venus was twelve or nine seconds before she resumed her rotundity."[3]

The phenomenon was estimated to last anywhere from a few seconds to a full minute; its duration seemed to depend as much on the observer's subjective impression as on any objective factors.

Various theories were proposed to explain the black drop effect: turbulence in Earth's atmosphere, an optical illusion, imperfections in the telescope, and the fallibility of the observer's own eyesight. It is well known, for example, that when a dark object is viewed against a bright background, the background light seems to engulf the object around its edge, making it appear smaller than it actually is. Figure 8.2 shows this effect at the moment of second contact. The thin lines in (a) indicate the true outlines of the sun and Venus just before immersion; but because of the sun's bright background, its edge seems to be receding outward to the black region, while that of Venus recedes inward. Thus at the true moment of immersion (b), instead of the two circles separating from each other internally, a thin ligament will still be seen to connect them for a few seconds.

To this optical illusion one must add the fact that our atmosphere is never completely calm. Turbulence in the air is always present, its effect being especially noticeable when a celestial object is viewed close to the horizon, when its rays of light must travel a greater distance through the atmosphere before reaching us (fig. 8.3). And finally, Venus's own atmosphere, discovered by Lomonosov during the 1761 transit, may also be a contributing factor.

(a)   (b)

FIGURE 8.2 Formation of the "black drop."

FIGURE 8.3 The effect of atmospheric distortion.

Whatever its cause, the black drop effect frustrated all attempts to record the exact moments of ingress and egress. Even worse, estimates of its duration differed widely even among observers at the same station, making any attempt to correct the timings by some agreed-upon factor totally futile. Thus, the great international effort to use the transit of 1761 to determine once and for all the solar distance ended in disappointment, if not outright failure.

But science, by its very nature, is an ongoing endeavor, and the failure of one attempt only spurs scientists to try again. A second chance was at hand just eight years hence. So the general attitude, after the 1761 transit results were analyzed, was to regard it as a dress rehearsal for the next transit, to take place on June 3, 1769. Surely the lessons learned would serve the observers well, and perhaps even the black drop effect could somehow be overcome. It was in this renewed spirit of optimism that preparations were

begun to observe the second and final transit of the eighteenth century.

***

THIS time the circumstances would be different. The 1761
transit was visible in its entirety from much of Asia and the
Indian Ocean; that of 1769 would be seen mainly from the
vast expanses of the Pacific Ocean. Only the Arctic region
and the western half of North America would be accessible
by land. A careful analysis of the circumstances showed that
the best observing locations would be in the southern Pacific, in a broad region midway between Australia and South
America—an area which at the time was almost completely
unknown. This, however, was not as great a drawback as
might be thought: in those days it was easier to reach a remote place by sea than over land—all the more so because
of the heavy equipment the observing teams had to carry
with them. Some governments even saw in the transit a pretext for exploring the South Seas, chart new waterways, and
perhaps even discover the much rumored *terra australis,*
the legendary southern continent whose existence had
been speculated for centuries. And lastly, the delicate time-
pieces which the expeditions would be carrying could be
tested and used in the ongoing quest to determine the longitude of a ship at sea.

The 1761 transit was thus barely over when plans were
already underway to observe the 1769 event. Many of the
same participants would again take part. Alexandre-Gui
Pingré was sent to the island of Santo Domingo in the
Caribbeans, where he successfully timed the moment of
ingress. The Abbé Chappe d'Auteroche, who had observed
the previous transit from the frigid wastes of Siberia, was

assigned to go to the Solomon Islands in the tropics, then under Spanish rule. But the Court of Spain, suspicious as always of foreigners, refused him permission, allowing him instead to sail with a Spanish fleet to Mexico. He arrived at the port of Veracruz, on the Gulf of Mexico, and from there crossed the country in an Indian-led safari to his destination, the mission of José del Cabo at the southern tip of Baja California. He arrived there on May 16, 1769, barely two weeks before the transit, and hastily set up his equipment. His observations were among the most successful anywhere, not only in timing the moments of ingress and egress, but also because he was able to determine his longitude with great precision, thanks to a superior pendulum clock. His mission accomplished, he was about to sail home when tragedy struck: an epidemic broke out in the village, claiming the lives of most of the inhabitants. All of Chappe's staff, save one, succumbed to the disease. Undeterred, he continued his astronomical observations for another month before falling victim himself on August 1, 1769; he was forty-one years old. The sole survivor, the geographer M. Pauly, had the sad task of bringing back Chappe's notes and instruments.[4]

✳✳✳

ENGLAND was less timid vis-à-vis the Spaniards: it went ahead with its plans without waiting for Spanish approval. In 1765 Thomas Hornsby, an astronomy professor at Oxford who had observed the 1761 transit, urged the Royal Society to draw up plans for the 1769 event. The Society, "unaccustomed to haste, proceeded in its usual leisurely manner."[5] A committee, headed by the fifth Astronomer Royal, Nevil Maskelyne, was formed to examine the matter

and recommend the most suitable observation sites. It first
met in November 1767, less than two years before the transit was due; after lengthy deliberations, it submitted its rec-
ommendations to the Society, which in turn drafted a memorandum to the king. It read in part:

> That the passage of the planet Venus over the disk of the
> sun, which will happen on 3rd of June in the year 1769,
> is a phenomenon that must, if the same be accurately ob-
> served in proper places, contribute greatly to the im-
> provement of Astronomy, on which Navigation so much
> depends.
>
> That several of the Great Powers in Europe, particu-
> larly the French, Spaniards, Danes and Swedes, are mak-
> ing the proper dispositions for the Observation thereof:
> and the Empress of Russia [Catherine the Great] has
> given directions for having the same observed in many
> different places of her extensive Dominions. . . .
>
> That the British nation has been justly celebrated in
> the learned world for their knowledge of astronomy, in
> which they are inferior to no nation upon earth, ancient
> or modern; and it would cast dishonour upon them
> should they neglect to have correct observations made of
> this important phenomenon. . . .
>
> That the expense of having the observations properly
> made . . . would amount to about £4,000, exclusive of the
> expense of the ship.

It added: "The Royal Society are in no condition to defray
this expense." The appeal to the king's sense of national
pride, as well as the reference to the practical benefits that
may result from the endeavor, had its intended effect: the
Society's request was readily approved.

Alexander Dalrymple, "a man eminent in science, a member of the Royal Society, and who has already greatly distinguished himself respecting the geography of the Southern Ocean,"[6] lobbied hard to be appointed as commander of the expedition. But the senior naval officer at the admiralty, Sir Edward Hawke, was adamantly opposed to appointing a civilian as commander of a naval vessel, declaring that "he would rather cut off his right hand than permit anyone but a King's Officer to command one of the ships of His Majesty's navy." No doubt Hawke had in mind Halley's experience when he was put in command of the ship that took him to St. Helena for the 1677 Mercury transit; the gentlemanly Halley, who commanded much respect among his friends and colleagues, had no experience in handling a bunch of unruly sailors.

After much wrangling and hand twisting, the admiralty finally agreed on a person fit for the task: Captain James Cook (1728–1779), who already had a distinguished naval career behind him. On May 25, 1768, Cook, then forty years old, was officially appointed as commander of the expedition and promoted to the rank of Lieutenant of the Royal Navy. His first task was to search for a vessel fit for the long voyage. "After examining a great number [of vessels] then lying in the Thames, they at last fixed upon the *Endeavour*, a barque of 370 tons which had been built for the coal-trade." She was thoroughly refurbished for her new task.

Cook's staff included Charles Green, who had been assistant to James Bradley, the third Astronomer Royal; he would be in charge of the astronomical aspects of the expedition. He in turn was assisted by Joseph Banks, Esq. (1743–1820), who would later become president of the Royal Society. "This friend of science possessed at an early

period of life an opulent fortune, and being zealous to apply
it to the best ends, embarked on this tedious and hazardous
enterprise, animated by the wish of improving himself and
enlarging the bounds of knowledge. He took two draughts-
men with him, and had likewise a secretary and four ser-
vants in his retinue." Banks used his enormous wealth to
fund numerous scientific projects, but he also distinguished
himself as a naturalist and botanist (Banks Peninsula, just
south of Christchurch, the capital of New Zealand's South
Island, is named after him).

There was one other scientist among the staff: "Dr.
Solander, an ingenious and learned Swede, who had been
appointed one of the librarians in the British Museum,
and who was particularly skilled as a disciple of Linnæus
[Carolus Linnæus (1707–1778), the renowned Swedish
botanist], and distinguished in his knowledge of natural history. Possessed with the enthusiasm with which Linnæus inspired his disciples, he braved danger in the prosecution of
his favourite studies; and being a man of erudition and ca-
pability, he added no small *éclat* to the voyage in which he
had embarked." One can gather from the makeup of Cook's
team that he had in mind more than just observing the transit. In fact, he was "directed, after he had accomplished his
main business, to proceed in making further discoveries in
the South Seas, which now began to be explored with uncommon resolution." Clearly, the admiralty was using the
transit as a pretext to advance its more ambitious goal of exploration and discovery.

The *Endeavour* raised anchor from Deptford on July 30,
1768, and after a brief stop in Plymouth, they were on their
way on the 26th of August. On April 10, 1769, after a long
but uneventful voyage, they reached their destination: the

island of Otaheite in the South Pacific, recently discovered
by Captain Samuel Wallis, who named it King George Island; today it is known as Tahiti. In May they began their
preparations for the transit. They built their main station at
Matavai Bay, on the northern tip of the island, and, not
knowing how they would be received by the natives, fitted
it as a fort as well as an observatory. As a precaution against
the vagaries of the local weather, they sent two additional
parties, one to the small islet of Puaru to the east, the other
to Tahiti's satellite island Eimeo (which the British called
York Island; it is known today as Moorea) to the west, where
they erected some tents and set up their telescopes. They
soon attracted the attention of the natives, who flocked
around the strange Europeans with their even stranger
paraphernalia. On April 15 Cook came ashore and officially
named the fort Point Venus, by which name it is still known
today.

The natives were generally friendly to the visitors and
watched intently their strange activities. But just days before the transit, Banks's heavy quadrant mysteriously disappeared, and they learned that it had been stolen by a native.
As this instrument was crucial for the transit observations,
a frantic search began at once to retrieve it. After an ardu-
ous pursuit up the steep hills in 90-plus degrees Fahren-
heit, they finally caught up with the thief and got back their
precious instrument, but the natives had meanwhile taken
it apart. In great haste it was reassembled, just in time for
the crucial event.

During the two weeks left before the transit, the weather
steadily deteriorated, with overcast skies and heavy showers occurring daily. At last the day arrived. Luck was with
them: "The morning of 3 June dawned crystal clear, as

favourable to our purpose as we could wish." Banks left his group to bring back some provisions; as he was negotiating with the natives, an unexpected visitor arrived: His Royal Highness King Tarras, the island's chief, accompanied by his sister Nuna. "Mr. Banks returned to the observatory with his visitors, and showed them the transit of the planet Venus over the sun's disk, informing them that he and his companions had come from their own country solely to view it in this situation." What was the natives' reaction to all this we are not told; surely they must have thought that these strangers from a different world were real oddballs.

The various parties watched the entire transit in splendid weather: "From horizon to horizon not a cloud was to be seen the whole day, and the air was perfectly clear; so that we had every advantage in observing the passage of Venus over the sun's disk." First contact was at 9 hours, 25 minutes, and 42 seconds A.M., with total immersion (second contact) nineteen minutes later. "Throughout that long morning, with the temperature rising to 119° F, the little black dot could be clearly discerned moving across the face of the sun until it touched the far side at 3 hours, 14 minutes and 8 seconds P.M."

And yet, Cook reported, "It was not until 9 a.m. that all those with eyes to the telescope recognised that, whatever the weather might be like over the Pacific, conditions were not so good on the surface of Venus, sixty-seven million miles distant, where the edges of the planet were obscured. We very distinctly saw an atmosphere, or dusky shade, round the body of the planet, which very much disturbed the times of contact, particularly the two internal ones." The black drop was once again playing havoc with the observers.

The transit was now over, and their official mission accomplished. But for Cook this was just the beginning. He continued his voyage to New Zealand (from where he watched the November 9, 1769, transit of Mercury), discovered the Admiralty and Society Islands, which he named in honor of the sponsors of his expedition, and landed in Australia, being the first Westerner to report in detail on the size and features of this unknown continent. In a second voyage (1772–1775) Cook reached Antarctica, and on his third and last voyage (1776–1779) he discovered the Hawaiian Islands; there, in the course of a dispute with the natives over a stolen boat, he was killed, and his body reportedly cannibalized. An obelisk marks the spot today.

✷✷✷

IF VENUS'S atmosphere had deprived Cook of complete success, at least he saw the entire transit under perfect conditions. Fate was not so kind to Le Gentil, who, as we recall, was waiting out the eight years between the two transits on the island of Mauritius. He crisscrossed the Indian Ocean from Madagascar to Manila, exploring the wildlife and natural history of the places he visited. But during all this time his thoughts were never far from the upcoming transit. He decided that Manila, capital of the Philippines, would be the best place from which to observe. Le Gentil duly informed the Academy of Sciences in Paris of his intentions, and asked it to get official permission from the Spanish authorities.

While his letter was on its way, a Spanish warship bound for Manila called port in Mauritius, and Le Gentil talked the captain into taking him aboard. He reached Manila on August 10, 1766, and at once began preparing for the transit. Then he received word from the Academy, instructing him to proceed to Pondicherry in India, his 1761 site, although only Venus's egress would be visible from there. He did as instructed and reached his old place in March 1768, still more than a year before the transit. He spent the remaining months preparing for the event, making all the necessary preliminary measurements, this time receiving generous help from the British.

Finally the crucial day arrived. On the night before, the sky was crystal clear, and Le Gentil confidently waited for the morrow. "On June 3, 1769, at the moment when this indefatigable observer was preparing to observe the transit, a vexatious cloud covered the sun, and caused the unhappy Le Gentil to lose the fruit of his patience and of his efforts."[7] He had missed the planet's egress, the moment when Venus was leaving the sun, not to return for a hundred and five years. As if to add to his frustration, he later learned that in Manila, on that day, the skies were perfectly clear! "It was two weeks before the ill-fated astronomer could hold the pen which was to tell his friends in Paris the story of his disappointment."[8]

But his troubles were not over yet. Having no chance to witness the next transit in 1874, he stayed in Pondicherry for a while, made some more observations, and slowly wound up his visit. At last he was ready to leave, having been away for almost twelve years. After being twice shipwrecked, he made it to Cadiz (Spain), and completed the last leg of his journey on foot across the Pyrenees. When he finally arrived in Paris, Le Gentil learned that he had been assumed dead by his heirs, who were busy dividing his estate. The Academy, not quite ready to sign him off yet, nevertheless demoted him to the rank of "veteran" (retiree), being convinced that he had neglected his official duties in order to make some personal gains. He was eventually given back his rank and position at the Paris observatory, but he had to take legal action to regain his personal property. The indefatigable Le Gentil then married, and spent the remaining twenty years of his life raising their daughter and writing up his papers. His major work, an account of his exploration of the Indian Ocean, was published in two volumes in 1779–1781. He died in 1792 at the age of sixty-seven.

✳✳✳

IN THE American colonies, then still under British rule, there was a great interest in observing the transit. This was partly in response to an appeal by Nevil Maskelyne, the Astronomer Royal, who, in a letter to Benjamin Franklin, had requested that an observing team be sent to Lake Superior, from where the entire transit would just be visible. But there was also an element of prestige involved: in Europe the upcoming transit had been trumpeted as one of the most significant events in the history of science, and the few scientists in the colonies were only too eager to prove they were up to par with their European colleagues. Several expeditions were organized in a rather improvised manner, as the colonies were short both in funds and in scientists capable of taking part.

John Winthrop, professor of mathematics at Harvard, had successfully observed the 1761 transit from New-foundland, but because of declining health he could not be counted on to take part in the 1769 event. It thus befell David Rittenhouse (1732–1796), regarded as the father of American astronomy, to take the leading role. Rittenhouse was born on a farm near Philadelphia and became a distin-

guished clock and instrument maker. An enthusiastic sup-
porter of the American Revolution, he served on the Safety
Committee in charge of the defense of Pennsylvania, and
in 1776 was instrumental in drafting the Revolutionary
Constitution of his state. But he is mainly remembered
today for his role in the 1769 transit, which he and two colleagues watched from an observatory built on his own farm.
They watched the progress of Venus in clear skies, but for
some reason Rittenhouse failed to record the exact moment
of ingress. According to one report, he briefly fainted at the
crucial moment; according to another, he was so overcome
by emotion at the first sight of Venus that his measurements
were adversely affected. He did notice, however, the faint
halo around the planet just before ingress, and he was successful in recording the moment of egress.[9]

Rittenhouse's observation, as those of many others along
the East Coast—nineteen of which were later published in
scientific journals—were highly praised in Europe; they
marked the beginning of America's rise to scientific promi-
nence. The event even left its mark on regional history: in
Providence, Rhode Island, a street near the hill from where
the transit was observed is still named Transit Street; from
atop this hill, at local noon on the day before the transit, a
cannon was fired, so that people could mark on their win-
dows the exact direction of the meridian—the north-south
line. It was no doubt an attempt to arouse public interest in
an event which meant little to the average citizen, but could
bring fame and money to the town.[10]

✷✷✷

ALL TOLD, the 1769 transit was watched by 151 official observers, and many more amateurs, at seventy-seven stations

spread over more than half the globe.[11] The solar parallax deduced from the combined volume of observations ranged from 8.50 to 8.88 arc seconds, with the most probable figure around 8.72; this was a definite improvement over the results from the previous transit, which spanned a much wider range, from 8.5 to 10.6 arc seconds. The new figures led to a solar distance between 92,049,650 and 96,162,840 miles.[12] Then in 1835, the results of both transits were reviewed by Johann Franz Encke (1791–1865), director of the Berlin Observatory, who put the sun's parallax at 8.571 arc seconds, with a margin of error of 0.037 arc seconds. His figure led to a solar distance of 95,370,000 miles; this figure, in the words of Simon Newcomb, became "a classic number adopted by astronomers everywhere, and familiar to every one who has read any work on astronomy."[13]

But the "classic number" would not last for long: the next two transits would prove it to be too high. Whereas all astronomers up to and including Kepler had grossly underestimated the sun's distance, now suddenly they were overestimating it! Nevertheless, the margin of error, though still considerably higher than Halley had hoped, was steadily narrowing.

Notes and Sources

1. I say "*some* standard meridian" because the designation of the meridian through Greenwich as the zero meridian was not universally accepted until 1913.

2. For a history of this problem, see Dava Sobel's excellent book, *Longitude* (New York: Walker and Co., 1995).

3. Richard A. Proctor, *Transits of Venus: A Popular Account of Past and Coming Transits* (1874; 4th ed. London: Longmans, Green, and Co., 1882), pp. 58–63.

4. The full story of this ill-fated expedition can be found in *The 1769 Transit of Venus: The Baja California Observations of Jean-Baptiste Chappe d'Auteroche, Vicente de Doz, and Joaquin Velázquez Cardenas de León.* Introduced and edited by Doyce B. Nunis, Jr. Translations by James Donahue, Maynard J. Geiger, O.F.M., and Iris Wilson Engstrand (Los Angeles: Natural History Museum of Los Angeles County, 1982).

5. This excerpt and the memorandum to the king that follows are taken from Richard Hough, *Captain James Cook: A Biography* (New York: W. W. Norton, 1994), pp. 36–37.

6. This and the subsequent quotations regarding Cook's expedition are from Proctor, *Transits of Venus,* pp. 75–80.

7. Ibid., p. 82, quoting M. Dubois in *The Passages of Venus in Front of the Disk of the Sun.*

8. Simon Newcomb, *Popular Astronomy* (New York: Harper, 1880), p. 182

9. See the article "America's Foremost Early Astronomer" by David Parry Rubincam and Milton Rubincam II, in *Sky & Telescope,* May 1995, on which this description is based.

10. For a full account of the American involvement in the 1761 and 1769 transits, see Brooke Hindle, *The Pursuit of Science in Revolutionary America: 1735–1789* (Chapel Hill: University of North Carolina Press, 1956), chapter 8.

11. They are listed in Harry Woolf, *The Transits of Venus: A Study in Eighteenth-Century Science* (Princeton, N.J.: Princeton University Press, 1959), pp. 182–187.

12. Proctor, *Transits of Venus,* pp. 85–92.

13. Newcomb, *Popular Astronomy,* p. 184.

9

The Next Two Appointments

> The semi-diameter of the earth's orbit is our standard measure of the universe. It is the great fundamental datum of astronomy—the unit of space, any error in the estimation of which is multiplied and repeated in a thousand different ways, both in the planetary and sidereal systems.
>
> Agnes M. Clerke, *A Popular History of Astronomy during the Nineteenth Century* (1885)

IT WOULD BE one hundred and five years before Venus appeared again on the sun's face. In those years, the world had changed once again. In the New World a new nation was born, the United States. In France, a bloody revolution toppled the monarchy and replaced it with a people's republic. Europe continued to wage its ethnic wars with unabated fury, all the while giving rise to intellectual and scientific achievements of the highest level. In 1781 Sir William Herschel (1738–1822) discovered a new planet, Uranus—the first expansion of the solar system since antiquity. On the opening night of the new century, January 1, 1801, Italian astronomer Giuseppe Piazzi (1746–1826) discovered the first of the minor planets, Ceres, to be followed by hundreds more in the coming years. These minor planets, or asteroids, neatly filled a mysterious gap between the orbits of

Mars and Jupiter, a gap whose existence seemed to defy a controversial "law," known as the Titius-Bode law, that supposedly governed the distances of the planets from the sun.[1] The year 1846 saw the sensational discovery of Neptune, whose existence had been predicted on purely theoretical grounds by Urbain Jean Joseph Leverrier (1811–1877) of France and John Couch Adams (1819–1892) of England; this feat marked the ultimate triumph of Newtonian mechanics and was hailed as one of the greatest achievements of the human mind.

Meanwhile, astronomers were shifting their attention to the universe beyond the solar system. In 1838 Friedrich Wilhelm Bessel (1784–1846) succeeded in measuring the minute parallax of a fixed star, 61 Cygni, caused by Earth's annual revolution around the sun. This allowed him to fix the star's distance at ten light years, or about 60 trillion miles from Earth—the first successful determination of the distance of a celestial object outside our solar system. New observational techniques were being introduced into astronomy, foremost among which were photography and spectroscopy. The first photographic images of the moon were taken in 1850, and soon the camera replaced the human eye as astronomy's chief tool. In 1859 Robert Wilhelm Bunsen (1811–1899) and Gustav Robert Kirchoff (1824–1887) invented the spectroscope, allowing scientists to unravel the physical composition of stars—the essence of astrophysics. Astronomy was advancing in leaps and bounds.

Some things, however, did not change. Venus, the goddess of love, the most brilliant object in the sky after the sun and moon, was still thought to be our twin planet, similar to Earth in every respect but for its proximity to the sun. And superficially the two planets do indeed show many similarities: Venus is just slightly smaller than Earth and just a little closer to the sun, and it has an atmosphere. The thick Venusian atmosphere shrouds the planet in an eternal blanket of clouds, and no one has ever seen any definite features on the planet's surface (although many have claimed as much). Since nothing was known, everything was possible, and astronomers went into wild speculations as to what is hidden under the heavy cloud cover. In their imagination they saw a planet teeming with life, if not intelligent life then certainly a rich tapestry of flora and fauna covering the planet from pole to pole. They even thought they could determine Venus's period of rotation—the length of its day: it had to be close to twenty-four hours, of course. Camille Flammarion (1842–1925), the great French popularizer of astronomy, wrote of Venus in 1880: "This world differs little from ours in volume, in weight, in density, and in the duration of its days and nights. . . . It should, then, be inhabited by vegetable, animal and human races but little different from those which people our planet." Forty years later, in 1923, his views were pretty much the same:

> Venus is a planet, a world like ours, and of the same size, which only shines by reflecting the sun's light into space. When one remembers that it is the same with us, and that, seen from the distance of some ten million miles, our Earth shines with a similar lustre, one is forced to admit that we are much more beautiful from afar than we are close by. . . . It is probably the same with Venus, so white and so radiant seen from here; possibly if we could go close to her, we should hear the cries of wild beasts in the forests, the battles of men devouring each other in so-

called civilised lands, and we might witness geological and human revolutions, more formidable on account of the fact that Venus, younger than ourselves, is less advanced in evolution.[2]

As for the planet's rotation period, he thought it was about twenty-four hours but was honest enough to admit that "this special point is not yet decided, and it is quite possible that the days are very long there, or rather, that they last for ever—perpetual day on one side and perpetual night on the other." Others were more specific. Johann Hieronymus Schröter (1745–1816), one of the most respected lunar observers of the late eighteenth century, put the rotation period squarely at 23 hours, 21 minutes, and 7.9 seconds. Just how far this figure was off the mark would become known only in the second half of the twentieth century.

On one thing, at least, everyone agreed: Venus has no satellites. If proximity and similarity to Earth demanded that our twin planet should have a moon, this was definitely disproved during the three transits observed thus far. Any satellite, even a small one, would have revealed itself starkly outlined against the sun's disk, but nothing of the sort was seen. By all available evidence—including images taken by spacecraft reaching the vicinity of Venus—none exists.[3]

***

THE importance of accurately determining the astronomical unit did not diminish since the transit of 1769. On the contrary, with astronomers shifting their attention to the fixed stars, this question became all-important. Bessel's determination of the parallax of 61 Cygni was only the first step; by the time of the 1874 transit, the distances of some twenty stars had been known with reasonable accuracy. But determining these distances depended on knowing the astronomical unit; the slightest inaccuracy in this unit would greatly affect the stellar distance scale. The solar distance thus became the yardstick of the universe. "The determination of this distance," wrote Simon Newcomb in 1880, "is one of the capital problems of astronomy."

A new factor also began to play a role—the general public. The sheer rarity of transits, glorified by the extraordinary adventures of those who had observed them in the past, was enough to fire the public's imagination. There was hardly a newspaper that did not report on the preparations to observe the next transit. Hayyim Selig Slonimski (1810–1904), a Jewish scholar who founded the first popular science journal in Hebrew, devoted an entire article to the upcoming event; using biblical phrases—Hebrew had not yet been adapted to modern usage—he explained to his readers why astronomers should care about such an odd event. In New York, the *Daily Graphic* ran a full-page story, accompanied by pictures of the Irish expedition posing in front of their telescopes; it explained that "the prime objective for which the transit is to be observed is to ascertain the distance of the earth from the sun, in which there is now an absolute error one way or the other (it is not certain which) of about five millions of miles" (fig. 9.1).

This unprecedented media coverage generated an enormous public interest that started a momentum of its own: European governments that a hundred years earlier had been reluctant to provide the means for sending even a modest observing team were now lavishing funds on their scientific institutions, not wanting to be seen as lagging behind their neighbors. To a modern ear it sounds familiar:

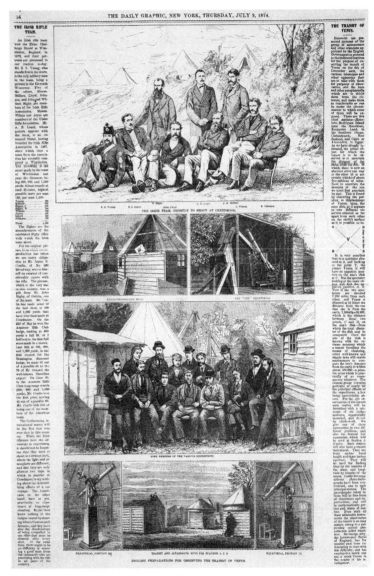

FIGURE 9.1 *Daily Graphic* (New York) article on the Irish Expedition of 1874.

the return of Halley's comet in 1986 generated a similar popular interest and prompted an international effort (in which the United States, because of cuts in NASA's budget, did not take part) to observe the famous cosmic visitor from a fleet of spacecraft sent to intercept it in its swing around the sun.

***

PREPARATIONS for the 1874 transit began well in advance; this time American astronomers took an active role, equal in every respect to their European colleagues. There was now not only the question of where to send the observing parties, but also the method of observation they should use. The new technique of photography raised everyone's hope that the notorious black drop effect could be avoided: surely a photographic image of the moment of internal contact would provide the exact instant of the event, devoid of any tricks played by the observer's eye. A special telescope, the photo-heliograph, was designed and built for the occasion. Its unique feature was that sun rays could be directed into it by an inclined mirror positioned at some distance from the telescope itself, thereby avoiding the need to aim the instrument directly at the sun. To ensure that all the participants were thoroughly trained for the occasion, the Naval Observatory in Washington, D.C., ran a series of practice drills, in which an "artificial transit" was created by having a black circle move slowly against a white background representing the sun (fig. 9.2). The apparatus was mounted on the top of Winder's Building near the War Department, and the observers practiced recording the moments of contact from the roof of the observatory, some 3,000 feet away.

FIGURE 9.2 An artificial transit.

As with all things of good intention, the question of who will pay for what inevitably came up. A request for $50,000 was presented to Congress and brought before its appropriations committee. Its chairman, James A. Garfield (soon to become America's twentieth president), was not sympathetic; he was won over only after the scientists appealed to his sense of national pride, arguing that all major European nations were sending their expeditions, and the United States should not be left behind. The money was finally granted; but as usually happens with such projects, the final budget far exceeded the original estimates. In the end, Congress appropriated $175,000 for the endeavor.

The day of the transit, December 9, was fast approaching, and the expeditions departed for their assigned stations around the globe. The weather played its usual tricks. After obtaining extensive meteorological data from the American consuls in China and Japan, it was decided to send one party to Peking (as it was then called) and another to Japan, the choice being between Nagasaki and Yokohama. After lengthy discussions about which of these locations offered the better chances, the choice fell on Nagasaki. As it turned out, on the crucial day the sky was perfectly clear in Yokohama but hazy in Nagasaki! A similar fate awaited the British party to the remote Kerguelen Island in the South Indian Ocean. Headed by the Reverend Joseph Perry

(1833–1889), director of the Stonyhurst Observatory near Manchester, it established its main observing station at the head of a deep inlet on the island's eastern shore, where prospects for clear skies seemed the best. Still, being worried about the constantly changing weather, Father Perry sent two smaller parties to nearby locations on the island. As it happened, the main observing party was totally clouded out at ingress but was able to record the moment of egress, while just the reverse happened to one of the subsidiary parties. The third party was successful at both contacts, as was a German group observing from the island's north coast.[4]

When the expeditions returned home, their results proved once again to be disappointing. The atmospheres of two planets, Venus and Earth, again conspired to render the observations imprecise, and even the photographic method could not entirely eliminate the black drop effect. The solar parallax derived from all the observations ranged from 8.79 to 8.83 arc seconds, corresponding to a solar distance ranging from 92,570,000 to 93,000,000 miles.[5] But this still left an uncertainty of 0.5 million miles—several times as large as astronomers had hoped. To quote the historian of astronomy Agnes M. Clerke: "As regards to the end for which it had been undertaken, the grand campaign had come to nothing."[6] It began to dawn on astronomers that they had put too much hope in the transit method, and that the ultimate determination of the astronomical unit would have to be sought elsewhere.

✱✱✱

INDEED, alternative methods had been tried for some time, and with encouraging results. Sir David Gill (1843–1914), a Scottish astronomer who observed the 1874 transit from Mauritius, became skeptical of its usefulness for determining the astronomical unit. He therefore "resolved to put the planet Mars on trial as a possible factor in solving the problem."[7] Building his own observatory on Ascension Island, in the mid-Atlantic Ocean, he made a series of excellent photographic measurements of the parallax of Mars in 1877, and later did similar work with some of the newly discovered asteroids. As a result of these measurements, the solar parallax was upgraded to 8.806 arc seconds, leading to a mean solar distance of 92,830,000 miles.[8]

It is not surprising, then, that there was considerably less enthusiasm for launching another major observing effort for the next transit, to occur on December 6, 1882. Still, the return of Venus to the sun's face was too rare an event to be passed over. In one of those bizarre twists that can only happen in the political arena, Congress was now quick to ap-

FIGURE 9.3 A medallion commemorating the 1874 and 1882 transits.

propriate $85,000 for a new observation program—this despite strong objections from several of the scientists themselves! Among those objecting was the Canadian-born Simon Newcomb (1835–1909). Newcomb began his career as a computer at the American Nautical Almanac Office in Washington ("computer" being the job description of a person of low scientific ranking who was assigned to do the drudgery of numerical calculations). He quickly rose to become professor of mathematics in the U.S. Navy and head of its observatory, and in this capacity was instrumental in coordinating the American expeditions for the 1874 transit. Even before that event, he had become doubtful of the usefulness of the transit method, and thought the money could be better spent on other projects. But once the money was appropriated, he agreed to head an expedition to the Cape of Good Hope in South Africa. He selected a station near the town of Wellington, some 40 miles northeast of Capetown, and watched the entire transit in perfect conditions. In his memoirs he wrote:

> On our departure we left two iron pillars, on which our apparatus for photographing the sun was mounted, firmly imbedded in the ground, as we had used them. Whether they will remain there until the transit of 2004, I do not know, but cannot help entertaining a sentimental wish that, when the time of that transit arrives, the phenomenon will be observed from the same station, and the pillars be found in such a condition that they can again be used.[9]

***

THE aftermath of the two nineteenth-century transits was anticlimactic, and interest in the method quickly faded. The task of comparing the numerous observations, and reducing them to a common database, required years of human calculations. Funds for this less glamorous phase of the endeavor were hard to get; according to Newcomb, a surplus of $3,000 from the original appropriation had been mysteriously spent on a different project and could not be retrieved. "The computers had therefore to be discharged and the work stopped until a new appropriation could be obtained from Congress." Then personal and legal feuds broke out among members of the commission in charge of the work, leading to Newcomb's resignation. In his memoirs, published in 1903, he wrote:

> I do not know that the commission was ever formally dissolved. Practically, however, its functions may be said to have terminated in the year 1886, when a provision of law was enacted by which all its property was turned to the Secretary of the Navy. What the present condition of the work may be, and how much of it is ready for the press, I cannot say. . . . Whether it will ever appear is a question for the future. All the men who took part in it or who understood its details are either dead or on the retired list, and it is difficult for one not familiar with it from the beginning to carry it to completion.

Recriminations were also the lot of the European astronomers. Having fallen short of their goal, scientists now had second thoughts about the wisdom of spending so much money on an endeavor which previous experience should have told them did not hold great promise of success. Their thoughts were expressed by Arthur R. Hinks, professor of astronomy at Gresham College in England, in his book *Astronomy:*

Unhappily, one must admit that the money spent on sending expeditions to Siberia, to the Pacific, to Kerguelen's Island in the Antarctic Ocean, was practically thrown away, through no fault of the astronomers, unless indeed they can be blamed for failure to foresee the difficulty which wrecked the whole enterprise. They had gone to determine the exact instant at which Venus appeared in contact with the edge of the sun's disc. They expected to have a perfectly clear-cut view of this contact, and they did not at all anticipate what actually happened, that the atmosphere round the planet, filled with sunlight, blurred the sharpness of the phenomenon and left it uncertain within a large number of seconds at what instant the critical phase took place.[10]

Thus a chapter in the history of astronomy came to an end. Five transits of Venus had been accurately predicted, and four of them meticulously observed by hundreds of astronomers around the globe, with the hope of gleaning from them a single number, the length of the astronomical unit. But Venus, the goddess of love and kindness, frustrated them all. They expected to see her bare figure starkly outlined against the sun's disk; instead, they saw her thinly veiled in a shroud of luminous light surrounding her body. When she returns again to the sun's face on June 8, 2004, she will be greeted by a different crowd, content to witness the rare show for its sheer beauty.

Notes and Sources

1. The law, first stated by Johann Daniel Titius (1729–1796) in 1766 and made popular by Johann Elert Bode (1747–1826) in 1772, says that if we write down the numbers 0, 3, 6, 12, 24, 48, . . . and add 4 to each,

we get the progression 4, 7, 10, 16, 28, 52, . . . , which represents the relative distances of the planets from the sun (Earth's distance being 10). The "law" is entirely empirical and has no theoretical basis whatsoever; nevertheless, it does give the correct planetary distances at least approximately, as the table (in which Earth's distance = 1) shows.

Planet	Distance According to Titius-Bode (Earth = 1)	Actual Distance (Earth = 1)
Mercury	0.4	0.39
Venus	0.7	0.72
Earth	1.0	1.00
Mars	1.6	1.52
Ceres	2.8	2.77
Jupiter	5.2	5.20
Saturn	10.0	9.54
Uranus	19.6	19.19
Neptune	38.8	30.07

Clearly, for most planets the fit is close, but for Neptune the discrepancy is quite large.

2. Camille Flammarion, *Dreams of an Astronomer,* trans. E. E. Fournier d'Albe (New York: D. Appleton, 1923), pp. 95 and 102–103.

3. An interesting account of an alleged discovery of a Venusian satellite is given by Joseph Ashbrook in his book, *The Astronomical Scrapbook: Skywatchers, Pioneers, and Seekers in Astronomy* (Cambridge, Mass.: Sky Publishing Corporation, and Cambridge, U.K.: Cambridge University Press, 1984), chapter 54.

4. For a fuller account of Father Perry's expedition, see ibid., chapter 44.

5. Simon Newcomb, *Popular Astronomy* (New York: Harper, 1880), pp. 202–203.

6. Agnes M. Clerke, *A Popular History of Astronomy during the Nineteenth Century* (1885; 3d ed. London: Adams and Charles Black, 1893), p. 292.

7. Mary Proctor, *Romance of the Sun* (New York and London: Harper, 1927), p. 53. Chapter 4 contains a fascinating account of Gill's work on Ascension Island as told by his wife, who served as his assistant. Mary Proctor was the daughter of Richard Proctor, author of *Transits of Venus;* she wrote several popular books on astronomy.

8. Gill, however, was not the first to use Mars for this purpose. Two hundred years before him, in 1672, Giovanni Domenico Cassini (1625–1712), director of the newly founded Paris Observatory, measured the position of Mars against the background stars during an exceptionally favorable opposition of the planet, while his assistant Jean Richer (1630–1696) did the same from the island of Cayenne, off the northwest coast of South America. By comparing their results, they determined the distance between the earth and sun to be 87 million miles. Though the precision of their measurements has recently been disputed, their finding nevertheless marks the first reasonably accurate determination of the astronomical unit. See Albert Van Helden, *Measuring the Universe: Cosmic Dimensions from Aristarchus to Halley* (Chicago and London: University of Chicago Press, 1985), chapters 12, 13, and 14.

9. Simon Newcomb, *Reminiscences of an Astronomer* (Boston and New York: Houghton, Mifflin & Co., 1903), p. 177.

10. As quoted in Mary Proctor, *Romance of the Sun*, p. 51.

Father Hell: Falsely Accused

THE ANNALS of astronomy are replete with controversial claims. Johann Hieronymus Schröter, who confidently put Venus's rotation period at 23 hours, 21 minutes, and 7.9 seconds, also claimed to have discovered a 28-mile-high mountain near Venus's south pole, a "discovery" that has never been confirmed. Others have claimed seeing a satellite around Venus. Most famous of all is Percival Lowell's announcement in 1895 that he detected "canals" on Mars, which he immediately attributed to the workings of an intelligent community, desperate to save their parched planet from global drought.

But it rarely happens that a *genuine* report is branded as false, misleading, or, worse, intentionally deceitful. Such was the case of Father Hell's observation of the 1769 transit. Maximilian Hell was born in Schemnitz, then in Hungary and now part of Slovakia, in 1720. In 1738 he entered the Jesuit order, and, following a long line of Jesuit scientists, embarked on a career in astronomy. In 1755 he became director of the newly founded University of Vienna Observatory, where, as we recall, he watched Venus's transit in 1761 in the presence of Archduke Joseph. His chief assignment at the observatory was to publish yearly ephemerides (tables of astronomical events), a task that occupied him for the next thirty-six years. His career and reputation thus seemed secure, so much so that King Christian VII of Denmark personally invited him to head an expedition to northern Norway (then part of Denmark) to observe the transit of 1769.[1]

Hell, accompanied by an assistant, left Vienna on April 28, 1768. They stopped in Lübeck (then part of Denmark and now in Germany), where they were received by the twenty-two-year-old king, and the three discussed Hell's plans in great detail. Then began the long journey northward over land and sea. They reached their destination, the island of Vardö off the northeast coast of Norway and well inside the arctic circle, on October 11, eight months before the transit was due. Vardö was the northernmost of the many stations from which the transit was to be observed; it was selected because the sun did not set there for much of the summer.

Hell chose his observing station near the small village of Vardöhus and began at once to build a frame observatory. He was helped by soldiers from the nearby royal fortress, but he complained about their "reporting for work at 9:30, take lunch from 11 to 2, and leave at 3:30, when it started to get dark."[2] Nevertheless, construction of the observatory, with its adjoining living quarters, was completed in time for the long arctic winter. Hell and his assistant then engaged in an extensive observation program, which included not only celestial sightings but also data about the geography, meteorology, and anthropology of this remote arctic region.

June 3, 1769, dawned clouded, but by an extraordinary stroke of luck the clouds dispersed just in time, and they recorded the moments of ingress and egress with great precision. To celebrate their success, nine salvos were fired from their ship's cannons, and they sang a hymn of praise to the Lord.

Now began a series of events that was to have unfortunate consequences. Hell was in no hurry to return to Vienna. He stayed in Vardö until June 27, making various additional observations, including a solar eclipse that happened to occur just one day after the transit.[3] He then slowly made his way home. En route he made plans to publish his many findings in a huge book, a plan that never came to fruition. But as far as his main mission was concerned—the transit observations—he was inexplicably silent. Time and again he delayed publishing his results, explaining that he wished to obey protocol and submit his work to the king before making it available to the public. His colleagues, however, became suspicious that he was delaying publication until the results from other stations would come in, so he could adjust his own to fit theirs; some even suspected that he did not have any data to report at all.

Hell then stopped in Copenhagen, where finally, on November 24, 1769, he submitted his report to the Danish Academy of Sciences. The Academy printed 120 copies for distribution among interested scientists, and for a while the controversy seemed settled. He returned to Vienna in August 1770, more than two years after his departure. He submitted his original notes to the Vienna Observatory and renewed his work there with undiminished energy, receiving many honors from his immediate colleagues. He died in 1792 at the age of seventy-one.

But suspicion about his transit observations lingered on. In 1835, Karl Ludwig Littrow, who would later become director of the Vienna Observatory, discovered Hell's original notes from Vardö, and found in them what appeared to be numerous corrections in ink. He immediately concluded that Hell had, in fact, altered his results so as to make them concur with those of other observers. So convinced was Littrow of his findings that he published them in a little book, which—as might be expected of a controversial work—received much attention. Hell's name was now tarnished, with no one to come to his defense.

Things stood thus for another fifty years. Then in 1883, Simon Newcomb visited the Vienna Observatory on official business. What he discovered there is best told in his own words:

In 1883 I paid a visit to Vienna for the purpose of examining the great telescope which had just been mounted there by Grubb, of Dublin. The weather was so unfavorable that it was necessary to remain two weeks, waiting for an opportunity to see the stars. One evening I visited the theatre to see Edwin Booth, in his celebrated tour over the Continent, play King Lear to the applauding Viennese. But evening amusements cannot be utilized to kill time during the day. Among the tasks I had projected was that of rediscussing all the observations made on the transits of Venus which occurred in 1761 and 1769, by the light of modern science. Hell's observations were among the most important made [due to Vardö's high latitude], if they were only genuine. So, during my almost daily visits to the observatory, I asked permission of Director Weiss to study Hell's manuscript.

At first the task of discovering anything which would lead to a positive decision on one side or the other seemed hopeless. To a cursory glance, the descriptions given by Littrow seemed to cover the ground so completely that no future student could turn his doubt into certainty. But when one looks leisurely at an interesting object, day after day, he continually sees more and more. Thus it was in the present case. One of the first things to strike me was that many of the alleged alterations had been made before the ink got dry. When the writer made a mistake, he had rubbed it out with his finger, and made a new entry.

The all-important point was a certain suspicious record which Littrow affirmed had been scraped out so that the new insertion could be made. As I studied these doubtful figures, day by day, light continually increased. Evidently the heavily written figures, which were legible, had been written over some other figures which were constantly beneath them, and were, of course, completely illegible, though portions of them protruded here and there outside of the heavy figures. Then I began to doubt if the paper had been scraped at all. To settle the question, I found a darkened room, into which the sun's rays could be admitted through an opening in the shutter, and held the paper in the sunlight in such a way that the only light which fell on it barely grazed the surface of the paper. Examining the sheet with a magnifying glass, I was able to see the original texture of the surface with all its hills and hollows. A single glance sufficed to show conclusively that no eraser had ever passed over the surface, which had remained untouched.

The true state of the case seemed to me almost beyond doubt. It frequently happened that the ink did not run freely from the pen, so that the words had sometimes to be written over again. When Hell first wrote down the little figures on which, as he might well suppose, future generations would have to base a very important astronomical element, he saw that they were not written with a distinctness corresponding to their importance. So he wrote them again over with the hand, and in the spirit of a man who was determined to leave no doubt on the subject, little weening that the act would give rise to a doubt which would endure for a century.

This, although the most important case of supposed alteration, was by no means the only one. Yet, to my eyes, all the seeming corrections in the journal were of the most innocent and commonplace kind,—such as any one may make in writing.

Then I began to compare the manuscript, page after page, with Littrow's printed description. It struck me as very curious that where the manuscript had been merely retouched with ink which was obviously the same as that used in the original writing, but looked a little darker than the original, Littrow described the ink as of a different color. In contrast with this, there was an important interlineation, which was evidently made with a different kind of ink, one that had almost a blue tinge by comparison; but in the description he [Littrow] makes no mention of this plain difference. I thought this so curious that I wrote in my notes as follows:—

"That Littrow, in arraying his proofs of Hell's forgery, should have failed to dwell upon the obvious difference between this ink and that with which the alterations were made leads me to suspect a defect in his sense of color."

Then it occurred to me to inquire whether, perhaps, such could have been the case. So I asked Director Weiss whether anything was known as to the normal character of Littrow's power of distinguishing colors. His answer was prompt and decisive. "Oh, yes, Littrow was color blind to red. He could not distinguish between the color of Aldebaran [the brightest star in the Bull] and that of the whitest star." No further research was necessary. For half a century the astronomical world had based an impression on the innocent but mistaken evidence of a color-blind man respecting the tints of ink in a manuscript.[4]

Father Hell, the veteran observer of the two eighteenth-century transits, was finally exonerated, thanks to the painstaking detective work of a veteran of the two transits of the nineteenth century. His honor was restored, and a large crater on the moon was named after him.[5]

Notes and Sources

1. The narrative that follows is based on three sources: Simon Newcomb, *Reminiscences of an Astronomer* (Boston and New York: Houghton, Mifflin and Co., 1903), pp. 154–160; Harry Woolf, *The*

Transits of Venus: A Study of Eighteenth-Century Science (Princeton, N.J.: Princeton University Press, 1959), pp. 176–179; and Joseph Ashbrook, *The Astronomical Scrapbook: Skywatchers, Pioneers, and Seekers in Astronomy* (Cambridge, Mass.: Sky Publishing Corporation, and Cambridge, U.K.: Cambridge University Press, 1984), chapter 43. These sources differ slightly in some details of Hell's expedition.

2. Ashbrook, *The Astronomical Scrapbook,* pp. 219–220.

3. This total eclipse occurred just six hours after the end of the transit. According to Patrick Poitevin's Solar Eclipse Calendar for the month of June, Venus should have been seen projected against the sun's corona, about one solar diameter from the edge of the solar disk. It must have been a spectacular sight, not to be repeated until June 6, 2263, when the next coronal transit (but not actual transit) will occur.

4. Newcomb, *Reminiscences of an Astronomer,* pp. 156–160.

5. See Hell's biography in the *Dictionary of Scientific Biography,* 16 vols., ed. Charles Coulston Gillispie (New York: Charles Scribner's Sons, 1970–1980). A brief biography of Hell can be found in Joseph MacDonnell, *Jesuit Geometers* (St. Louis: Institute of Jesuit Sources, and Vatican City: Vatican Observatory Publications, 1989, p. A-7). MacDonnell claims that Hell "fell victim to the public defamation of Jesuits then in vogue when he was accused of altering his findings during a transit of Venus."

Afterthoughts

MANY PEOPLE'S image of a scientist is that of a cold, unemotional person, whose rational thinking allows no display of feelings. That may well be the image some scientists would like to project of themselves, but it hardly fits reality. To prove the point, let me bring here excerpts from Sir Robert Staywell Ball's description of his observation of the transit on December 6, 1882, from Dunsink Observatory near Dublin. Ball (1840–1913) was born in Dublin, son to the secretary of the municipal zoo. He studied mathematical physics at Trinity College in Dublin, and in 1865 became tutor to the son of William Parsons, third Earl of Rosse. The elder Rosse made himself a name as builder of the largest telescope then in existence, a 58-foot-long behemoth whose 72-inch mirror weighed four tons (its construction was completed by his son, and the telescope was erected near the present town of Birr in Ireland).[1] In 1874 Ball became Astronomer Royal of Ireland, and in 1890 he moved to Cambridge to become director of the observatory there. His main research was in mathematical astronomy, but he is chiefly remembered for his many popular books and lectures on astronomy. A contemporary, H. Montagu Butler, described him as "a lecturer whose science and wit and playfulness combined can absolutely rivet any audience, from a savant to a little child." The following excerpts are from Ball's immensely popular *The Story of the Heavens* (London: Cassell, 1886).

The morning of the eventful day appeared to be about as unfavourable for a grand astronomical spectacle as could well be imagined. Snow, a couple of inches thick, covered the ground, and more was falling, with but little intermission, all the forenoon. It seemed almost hopeless that a view of the phenomenon could be obtained from that observatory; but it is well in such cases to bear in mind the injunction given to the observers on a celebrated eclipse expedition. They were instructed, no matter what the day should be like, that they were to make all their preparations precisely as they would have done were the sun shining with undimmed splendour. By this advice no doubt many observers have profited; and we acted upon it with very considerable success.

There were at that time at the observatory two equatorials, one of them an old, but tolerably good, instrument, of about six inches aperture; the other the great South equatorial, of twelve inches aperture. At eleven o'clock the day looked worse than ever; but we at once proceeded to make all ready. I stationed Mr. Rambaut at the small equatorial, while I myself took charge of the South instrument. The snow was still falling when the domes were opened; but, according to our prearranged scheme, the telescopes were directed, not indeed upon the sun, but to the

place where we knew the sun was, and the clockwork was set in motion which carried round the telescopes, still constantly pointing towards the invisible
sun. The predicted time of the transit had not yet
arrived. . . .

Up to one o'clock not a trace of the sun could be
seen. Shortly after one o'clock, however, we noticed
that the day was getting lighter; and, on looking to the
north, whence the wind and the snow were coming,
we saw, to our inexpressible delight, that the clouds
were clearing. At length, the sky towards the south
began to improve, and at last, as the critical moment
approached, we could detect the spot where the sun
was becoming visible. But the predicted moment arrived and passed, and still the sun had not broken
through the clouds, though every moment the certainty that it would do so became more apparent. The
external contact was therefore missed. We tried to
console ourselves by the reflection that this was not,
after all, a very important phase, and hoped that the
internal contact would be more successful.

At length the struggling beams pierced the ob-
struction, and I saw the round, sharp disk of the sun
in the finder, and eagerly glanced at the point on
which attention was concentrated. Some minutes
had now elapsed since the predicted moment of first
contact, and, to my delight, I saw the small notch in
the margin of the sun showing that the transit had
commenced, and that the planet was then one-third
on the sun. But the critical moment had not yet arrived. . . . This first [internal] contact was timed to
occur twenty-one minutes later than the external

contact. But the clouds again disappointed our hope. . . .

While steadily looking at the exquisitely beautiful sight of the gradual advance of the planet, I became aware that there were other objects besides Venus between me and the sun. They were the snowflakes, which again began to fall rapidly. I must admit the phenomenon was singularly beautiful. The telescopic effect of a snowstorm with the sun as a background I had never before seen. It reminded me of the golden rain which is sometimes seen falling from a flight of sky-rockets during pyrotechnic displays; but I would gladly have dispensed with the spectacle, for it necessarily followed that the sun and Venus again disappeared from view. The clouds gathered, the snowstorm descended as heavily as ever, and we hardly dared to hope that we should see anything more; 1 hr. 57 min. came and passed, the first internal contact was over, and Venus had fully entered on the sun. We had only obtained a brief view, and we had not yet been able to make any measurements or other observations that could be of service. Still, to have seen even a part of a transit of Venus is an event to remember for a lifetime, and we felt more delight than can be easily expressed at even this slight gleam of success.

But better things were in store. My assistant came over with the report that he had also been successful in seeing Venus in the same phase as I had. We both resumed our posts, and at half-past two the clouds began to disperse, and the prospect of seeing the sun began to improve. It was now no question of the observations of contacts. Venus by this time was well on the sun, and we therefore prepared to make observations with the micrometer attached to the eyepiece. The clouds at length dispersed, and at this time Venus had so completely entered on the sun that the distance from the edge of the planet to the edge of the sun was about twice the diameter of the planet. . . . The sun was now very low, and the edges of the sun and of Venus were by no means of that steady character which is suitable for micrometrical measurement. The margin of the luminary was quivering, and Venus, though no doubt it was sometimes circular, was very often distorted to such a degree as to make the measures very uncertain.

We succeeded in obtaining sixteen measures altogether; but the sun was now getting low, the clouds began again to interfere, and we saw that the pursuit of the transit must be left to the thousands of astronomers in happier climes who had been eagerly awaiting it. But before the phenomenon had ceased, I spared a few minutes from the somewhat mechanical work at the micrometer to take a view of the transit in the more picturesque form which the large field of the finder presents. The sun was already beginning to put on the ruddy hues of sunset, and there, far in on its face, was the sharp, round, black disk of Venus. It was then easy to sympathise with the supreme joy of Horrocks, when, in 1639, he for the first time witnessed this spectacle. The intrinsic interest of the phenomenon, its rarity, the fulfillment of the prediction, the noble problem which the transit of Venus helps us to solve, are all present to our thoughts when we look at this pleasing picture, a repetition of which will not occur again until the flowers are blooming in the June of A.D. 2004.

Notes and Sources

1. This telescope has recently been restored to its original grandeur; see the article "The Leviathan Reborn" by Patrick Moore, *Sky & Telescope,* November 1997, pp. 52–54.

10

Transits of Fancy

> I saw, as one might see the transit of Venus, a quantity
> passing through infinity and changing its sign from plus
> to minus. I saw exactly how it happened . . . but it was
> after dinner and I let it go.
>
> Winston Churchill, *My Early Life* (1930)

SINCE INTELLIGENT life first evolved on our planet, Venus
has transited the sun thousands of times, but only five have
been witnessed by humans. This, at least, is the evidence as
we know it from confirmed records. Were it not for the
blinding brilliance of the sun, a transit of Venus could be
seen with the unaided eye. We cannot therefore rule out the
possibility that some ancient observer with a keen eye,
watching the sun on a particularly hazy day, might have
glimpsed Venus's dark image on the solar disk. There have
indeed been occasional reports of such sightings. Camille
Flammarion tells us that "thirty-five centuries ago the
Babylonians observed one of [Venus's] transits across the
sun."[1] The *Norton History of Astronomy and Cosmology*
mentions a transit allegedly seen by the Aztecs:

> The worship of Venus was no less universal and im-
> portant than that of the sun, and gave rise to laudable
> astronomical techniques of prediction. The planet was

regularly and closely observed. There used to be a picture—now destroyed by the tourists' habit of throwing bottles at it—near a sacred underworld lake at Chich'en Itza, a lake into which sacrificial victims were thrown. This showed a square sun rising over the horizon, and it carried a date equivalent to 15 December 1145. Modern calculations show that a rare transit of Venus across the face of the sun was indeed to be seen on that day.[2]

If true, this would indeed have been an extraordinary event; but a quick check with Jean Meeus, one of the world's leading authorities on dating past and future astronomical events, shows that no transit occurred on that day. According to Meeus, the closest transit to the alleged date occurred on November 23, 1153.[3] It should also be noted that ordinary sunspots—which can occasionally be seen with the unaided eye during a hazy sunrise or sunset— have often been mistaken for transits. Nevertheless, to quote Meeus, "It is not impossible that a transit of Venus has been seen by chance in ancient times, near sunrise or sunset."

On the other hand, transits have occasionally been reported which occurred only in the mind of the reporter. Inexperienced amateurs have often been quick to report strange objects passing across the sun or moon; these, most likely, were simply a passing bird, a distant hot-air balloon, or a sunspot mistaken for a transiting planet. Simon Newcomb, always on the watch to debunk astronomical misinformation, tells of an incident in which a telescopic observation was announced "of something which we now know must have been flights of distant birds over the disk of the sun." This non-event was published in a leading journal, the *Astronomische Nachrichten* (Astronomical News) as "a wonderful transit of meteors."[4]

Perhaps the most famous of these ghost transits is the case of French amateur astronomer Edmond Modeste Lescarbault (1814–1894).[5] A physician by profession, he built himself an observatory at his home in the small village of Orgères-en-Beauce, some 50 miles southwest of Paris. With his modest $3\frac{3}{4}$-inch telescope, he watched the sun whenever he could spare a moment from his medical duties. In 1845 he had observed a transit of Mercury, an event which, though not exactly spectacular, must have left a deep impression on him. Fifteen years later the memory of that event came back to him when, on March 26, 1859, he noticed a small black round spot on the sun's face near its edge. Being familiar with solar observations, Lescarbault at first took it to be a sunspot. But the object showed a distinct movement, and he knew he had seen something else. He took some rough measurements of its position and direction of motion, but just then he was called to duty by a patient who required immediate attention. When he returned to his telescope the object was still there, but by then it had almost reached the sun's limb. Using a primitive clock that could read only minutes but not seconds, he estimated the total duration of the transit at 1 hour, 17 minutes, and 9 seconds, the seconds being deduced from the swings of a pendulum with which he took his patients' pulse. His object had barely grazed the solar disk, but that was enough to cause a stir in the astronomical community.

Being of a shy disposition, Lescarbault did not report his observation until nine months later, when he read in a science journal an article on a hypothetical planet assumed to orbit the sun *inside* Mercury's orbit. This putative planet had been proposed by renowned French astronomer Urbain Jean Joseph Leverrier (1811–1877) in an attempt to account for a mysterious anomaly in Mercury's orbit. In 1845 Leverrier had explained an annoying irregularity in the orbit of Uranus by assuming the existence of a transuranian planet which, by its gravitational pull on Uranus, perturbs the latter from its "normal" orbit. Using only paper and pen and Newton's theory of gravitation, Leverrier calculated the position and time where the new planet should be seen. His moment of triumph came on the night of September 23, 1846, when the new planet, later to be called Neptune, was discovered at the Berlin Observatory by Johann Gottfried Galle. It was a stunning triumph for Newtonian mechanics and for Leverrier personally; he became a world celebrity.[6]

The discovery of Neptune solved the problem of Uranus's irregular motion, at least for a while. But it was a different story with Mercury. Its highly eccentric orbit, and the fact that it never strays away from the sun more than 28 degrees as seen from Earth, conspired to make its orbit the least understood among the known planets. To complicate matters even further, there was a tiny residual displacement in Mercury's orbit that could not be explained by any known cause. Kepler's first law says that each planet moves around the sun in an ellipse. Had there been only a single planet orbiting the sun, its elliptical orbit would be fixed in space relative to the fixed stars; that is, the ellipse's long axis would keep the same direction in space. But this is an idealized picture. There are not one but nine major planets circling the sun, each exerting a gravitational pull on all the others. As a result, the long axis of each planet's elliptical orbit performs a slow rotation of its own. To be sure, this residual motion, or *precession of the perihelion* (perihelion = the orbit's closest point to the sun), is very small and can be accounted for by the combined gravitational perturbations caused by the other planets—in particular, Jupiter and Saturn. In the case of Mercury, however, there remained a residual motion of 43 arc seconds *per century* that could not be accounted for even after the perturbations caused by all the other planets had been taken into consideration. This minute, unaccounted motion of Mercury became the bane of nineteenth-century astronomy.

Leverrier, flushed with glory after predicting the existence of Neptune, now applied the same reasoning to Mercury: he attributed the unaccounted motion to the gravitational pull of an unknown intermercurial planet. And because its orbit was inside Earth's, this planet, like Mercury and Venus, ought occasionally to pass in front of the sun. Then, in December 1859, he received a letter from an unknown amateur astronomer, claiming that he, the amateur, had indeed seen the new planet pass in front of the sun. The author of the letter was Lescarbault, who, after reading about Leverrier's theory, decided it was time to inform him of what he had seen nine months earlier.

Leverrier jumped at the news as if he had hit the jackpot. He immediately took the train to Orgères, where Lescarbault resided, and, unannounced, knocked on the doctor's door. Not wasting time to introduce himself, he bombarded the frightened Lescarbault with a barrage of questions on the technical details of his sighting. Lescarbault answered as best as he could, and after an hour of intense interrogation, Leverrier was satisfied that his hypothetical planet had, indeed, been found. Rushing back to Paris, he attended a meeting of the French Academy of Sciences on

January 2, 1860, at which the discovery of the intermercurial planet was announced with great fanfare. Lescarbault,
the timid amateur, was praised as a scrupulous man, whose
"observations ought to be admitted to science." In England,
the Royal Astronomical Society recognized "the singular
merit of M. Lescarbault's observations. . . . Astronomers of
all countries will unite in applauding this second tri-
umphant conclusion to the theoretical inquiries of M. Leverrier."[7] Lescarbault was given the Legion of Honor and
was showered with requests to appear before Paris's
learned societies; he declined them all.

But it was Leverrier who got the lion's share of the acco-
lade. His fame skyrocketed, and he was hailed as second
only to Newton. What's more, unlike the discovery of Neptune, this time it was a French discovery through and
through: a renowned French scholar and aristocrat had predicted the existence of a new planet, and an unknown
French country doctor confirmed his prediction in a single
observation. Paris was agog with excitement. And the new
planet? It was given the name Vulcan, the Roman god of
fire, as befits a planet so close to the sun.

As with any major news story, there were those who used
the discovery to serve their own agenda. One leading American newspaper berated astronomers at the Harvard Observatory at Cambridge for having "permitted an amateur
astronomer, armed with a telescope no more powerful than
an opera glass, to make this greatest of all modern discoveries in the heavens. . . . These gentlemen at the Cambridge Observatory, having at their command the best instruments in America, seem to have been asleep or too lazy
to do anything except to draw their pay."[8] Harsh words indeed, and premature, too.

Based on Lescarbault's single sighting, Leverrier now computed Vulcan's orbit: it circled the sun in a near circular orbit at 13 million miles, or about 0.147 astronomical units. The period of revolution was 19 days and 17 hours, and Vulcan's orbit was inclined to the ecliptic (the plane of Earth's orbit) by 12 degrees.

As soon as the existence of Vulcan was announced, other amateurs, and some professionals as well, suddenly remembered having seen the planet transit the sun before. Few of these reports were reliable; some referred to observations made many years earlier, and some couldn't even quote the exact date when they were made. Nevertheless, Leverrier changed Vulcan's orbital elements with each new report. He even announced the probable dates of future Vulcan transits, and astronomers worldwide were intently awaiting them. But they waited in vain: no transit of any kind was seen. It seems incredible that a man of Leverrier's standing should ignore one of the iron rules of observational astronomy: that no discovery is given any credence until independently verified by other observers. He had been so obsessed with his hypothetical planet that he put all his faith in Lescarbault's single report.

What exactly Lescarbault saw on March 26, 1859, we shall never know. But we do know this: planet Vulcan was a fiction. It was extensively searched for during several total solar eclipses, when even a small planet near the sun should have been visible, if not to the eye then certainly to a sensitive photographic plate. Nothing whatsoever was found. To quote Newcomb:

During the past fifty years, the sun has been observed almost every day with the greatest assiduity by eminent observers, armed with powerful instruments, who have made the study of the sun's surface and spots the principal work of their lives. None of these observers has ever recorded the transit of an unknown planet. This evidence, though negative in form, is, under the circumstances, conclusive against the existence of such a planet of such magnitude as to be visible in transit with ordinary instruments.[9]

This was said in 1883. A century later the same conclusion still holds: none of the dozen or so spacecraft launched to the vicinity of Mercury and Venus returned any images of an unknown body.

Lescarbault lived long enough to witness the transit of Venus on December 6, 1882, which he watched from the same place where, twenty-three years earlier, he had made his "observation" of Vulcan. He died twelve years later, his name all but forgotten.

Leverrier's fame, justly resting on his prediction of Neptune's existence, has outlived him. He died in 1877, honored and remembered with great admiration. But the mystery of Mercury's residual motion, those unaccounted 43 arc seconds per century, remained unsolved for another half a century. All kinds of explanations were offered, among them the proposition that not just a single planet, but a cluster of small planetoids circles the sun inside Mercury's orbit. Even more bizarre was the suggestion that the exponent in Newton's law of gravitation is not 2 (whence the name "inverse square law"), but slightly greater than 2. These arbitrary hypotheses were invented solely in order to rescue Mercury from its tiny unaccounted residual motion; they had no physical basis whatsoever, and their introduction caused more problems than they solved.

The nineteenth century was coming to a close, and with it classical physics. The year 1900 signaled the birth of modern physics, when Max Planck (1858–1947) postulated that energy is not continuous, but consists of tiny bits which he called *quanta*. Five years later, Albert Einstein (1879–1955) published his special theory of relativity and revolutionized the way we perceive space and time. He then set his aim even higher: he attempted to revamp the concept of gravitation itself. The result was his general theory of relativity, hailed as the most esthetically satisfying physical theory ever created. Among other things, it said that light should be bent in the presence of a strong gravitational field, such as that in the sun's vicinity.

Einstein knew that, with all its esthetic appeal, the ultimate test of his theory will lie in its experimental verification. He proposed three tests that could be performed, at least in principle—all of them in the realm of astronomy. The first of these involved the infamous residual motion of Mercury. In November 1915, as he was giving his theory its final touches, he applied his equations to Mercury. To his unbounded joy, they showed that Mercury's perihelion should advance at a rate of 43 arc seconds per century—exactly the amount that had given astronomers so much trouble. Einstein was beside himself with excitement: "I knew that nature is telling me something." Vulcan lost its raison d'être and soon vanished into oblivion.[10]

Notes and Sources

1. Camille Flammarion, *Dreams of an Astronomer*, trans. E. E. Fournier D'Albe (New York: D. Appleton, 1923), p. 95.

2. By John North (New York: W. W. Norton, 1995), pp. 156–157.

3. Jean Meeus, *Transits* (Richmond, Va.: Willmann-Bell, 1989), p. 47. The date is by the Julian or "old style" calendar, hence its occurrence in November rather than December.

4. Simon Newcomb, *Popular Astronomy* (New York: Harper, 1880), p. 294. A spectacular photo showing the International Space Station transiting the sun during the partial eclipse of December 25, 2000 (the "Christmas Eclipse" and the last solar eclipse of the second millennium), appeared in *Sky & Telescope*, April 2001, p. 124.

5. The story that follows in based on Richard Baum and William Sheehan, *In Search of Planet Vulcan: The Ghost in Newton's Clockwork Universe* (New York and London: Plenum Trade, 1997), chapter 10.

6. In England, John Couch Adams (1819–1892) arrived at the same prediction independently of Leverrier. See Morton Grosser, *The Discovery of Neptune* (1962; rpt. New York: Dover, 1979).

7. Baum and Sheehan, *In Search of Planet Vulcan*, p. 155.

8. Ibid., pp. 155–156.

9. Simon Newcomb and Edward S. Holden, *Astronomy*, 3d ed. (New York: Henry Holt, 1887), p. 227.

10. Nevertheless, isolated attempts to search for intermercurial objects have been made from time to time; see Baum and Sheehan, *In Search of Planet Vulcan*, pp. 253–254.

11

A View from Other Worlds

> Although such events [transits seen from other planets] cannot be observed from our earthly world and hence have—at least for the present time—only a theoretical interest, their consideration can be enlightening. . . . And is curiosity not the mother of Science?
>
> Jean Meeus, *Transits* (1987)

"LOCATION, location, location" is the realtor's credo. Add to it "Timing, timing, timing," and it becomes the astronomer's credo as well, for everything in astronomy depends on being at the right place at the right time. On October 27, 1780, a total solar eclipse was visible from the state of Maine, and Harvard College sent an expedition to watch it from Penobscot Bay. Notwithstanding the Revolutionary War, the group made it safely to its assigned location, set up its equipment, and eagerly awaited the event. The skies were clear and the eclipse commenced at the predicted time, but the moment of totality never arrived. The moon, instead of completely blocking the sun's rays, left a sliver of the solar disk uncovered. None of the awe-inspiring spectacles of totality were visible: no corona, no solar flares, no stars suddenly visible at noon. To their dismay, they found out that they had positioned themselves a few miles *outside* the path of totality. What had gone wrong has been disputed ever since; the expedition's leader blamed it on inaccurate maps, while the leader himself was blamed by his colleagues for reading the maps wrongly.[1]

As long as we humans are confined to the earth beneath our feet, our choice of location from which to observe a celestial event is limited to the surface of our home planet. But the day will come, perhaps sooner than we imagine, when humans will travel to other planets. New vistas will then open up that are forever denied to us earthlings. An astronaut on Mars, for example, might witness not only a transit of Mercury or Venus, but of our own earth. A spectacular sight it must be: a small black circle slowly encroaching on the solar disk, to be followed some six hours later by a second, still smaller black circle—our own moon. As Earth's image makes its way from east to west across the solar disk, our astronaut will actually see the moon's position change in relation to the earth, just as Galileo saw the four satellites of Jupiter change their position in his telescope. But this time the entire spectacle could be seen with the unaided eye, and not just during a transit: the moon, about one-fourth of Earth's diameter, would easily be visible in the Martian night sky as a bright "star" moving back and forth around a much brighter Earth. Had our ancestors lived on Mars, perhaps it would have been easier for them to accept the fact that in the universe, small bodies move around large ones—the essence of the Copernican system.

Jean Meeus, the dean of astronomical computing (b. 1928 in Belgium), has made it his specialty to date odd celestial events that have never been seen, and probably never will be. According to Meeus, two Earth transits were visible from Mars in the twentieth century: on May 8, 1905, and on May 11, 1984.[2] The next such event will not happen until November 10, 2084, by which time humans may have already set foot on the red planet. Inspired by Meeus's findings, the science fiction writer Arthur C. Clarke wrote a moving story, "Transit of Earth," in which a lonely astronaut, stranded on Mars because his spaceship had been damaged during landing, spends his last hours watching his home planet transit the sun:

> The radio has just printed out a message from Earth, reminding me that the transit starts in two hours. As if I'm likely to forget—when four men have already died so that I can be the first human being to see it. And the only one, for exactly a hundred years. It isn't often that Sun, Earth, and Mars line up neatly like this; the last time was in 1905, when poor old Lowell was still writing his beautiful nonsense about the canals and the great dying civilization that had built them. Too bad it was all delusion. . . .
>
> There are only five minutes to go. All the equipment is in perfect condition. The telescope is tracking the sun, the video recorder is standing by, the precision timer is running. . . . These observations will be as accurate as I can make them. I owe it to my lost comrades, whom I'll soon be joining. They gave me their oxygen, so that I can still be alive at this moment. . . .
>
> Only a minute to go; getting down to business. For the record: year, 1984; month, May; day, 11, coming up to four hours thirty minutes Ephemeris Time . . . *now*.
>
> Half a minute to contact. Switching recorder and timer to high speed. Just rechecked position angle to make sure I'm looking at the right spot on the sun's limb. Using power of five hundred—image perfectly steady even at this low elevation.

Four thirty-two. Any moment now. . . . There it is . . . there it is! I can hardly believe it! A tiny black dent in the edge of the Sun . . . growing, growing, growing. . . . Hello, Earth. Look up at me, the brightest star in your sky, straight overhead at midnight. . . .

Now I can see the effects of the atmosphere. There's a thin halo of light surrounding that black hole in the Sun. Strange to think that I'm seeing the glow of all the sunsets—and all the sunrises—that are taking place around the whole Earth at this very moment. . . .

Ingress complete—four hours fifty minutes five seconds. The whole world has moved onto the face of the Sun. A perfectly circular black disk silhouetted against that inferno ninety million miles below. It looks bigger than I expected; one could easily mistake it for a fair-sized sunspot.

Nothing more to see now for six hours, when the Moon appears, trailing Earth by half the Sun's width. I'll beam the recorder back to Lunacom [lunar communication base], then try to get some sleep. . . .

Back at the telescope. Now the Earth's halfway across the disc, passing well north of center. In ten minutes, I should see the Moon. Damn—missed it. Doesn't matter—the recorder will have caught the exact moment. There's a little black notch already in the side of the sun. First contact must have been about ten hours forty-one minutes twenty seconds ET. . . .

Ten hours fifty minutes, Recorder off. That's it—until the end of Earth transit, two hours from now.[3]

Transits of Earth as seen from Mars are more frequent than those of Venus as seen from Earth, but their cycle is less regular. They often repeat after seventy-nine years (which is equal to forty-two Mars years), but this cycle is not permanent: after the 1905–1984 pair (fig. 11.1) there is a transit-free period of one hundred years until the next transit in 2084. Occasionally a cycle may contain three transits, as will happen in 2394, 2473, and 2552 (fig. 11.2); by contrast, the transits of 927 and 948 were separated by just twenty-one years. At present, all Martian Earth transits occur in May or in November.

The farther out we go in the solar system, the more planets will be left inside, each with its own transit schedule. From Jupiter, for example, an observer may see transits of Mercury, Venus, Earth and Mars, as well as those of hundreds of asteroids; Jupiter itself is added to the list if we could be standing on Saturn. It is a peculiar fact that although the sun will steadily diminish in size as we go farther out, the duration of a transit will generally *increase*—

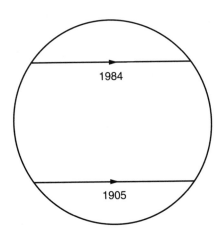

Figure 11.1 Earth transits as seen from Mars, 1905 and 1984.

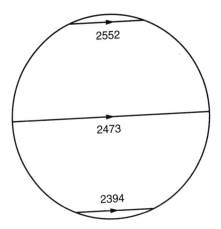

FIGURE 11.2 Earth transits as seen
from Mars, 2394, 2473, and 2552.

a consequence of Kepler's third law. For example, as seen
from Mars, the sun has a mean diameter of 1260 arc seconds, compared to 1919 arc seconds from Earth; but an
Earth transit as seen from Mars can last up to 9.5 hours,
compared to about 8 hours for a Venus transit from Earth
(these numbers refer to *central transits,* when the path of
the planet crosses the center of the solar disk). Excluding
Pluto, the longest possible transit visible from anywhere in
the solar system is that of Uranus as seen from Neptune: almost 42 hours. On the other hand, the planet that covers
the greatest *proportion of the solar disk* during a transit is
Jupiter as seen from Saturn: the giant planet will be eclipsing more than one-fifth of the solar diameter, itself a mere
200 arc seconds wide. But Jupiter will not be the only body
seen on the solar disk: it will be accompanied by its four
large moons and at least a dozen smaller companions clus-
tering around it like a pack of loyal watchdogs. Appendix 3

has details of some upcoming transits as seen from various planets.[4]

Meeus has also dated some of the rarest astronomical events to be seen from our own planet. Have Mercury or Venus ever been in transit *during a solar eclipse?* Meeus has found that no such event occurred between the years 0 and 3050; but there was a close call during the total eclipse of June 4, 1769, which occurred just a few hours after the end of Venus's transit on the June 3. Had the two events actually coincided, it would have been an occasion of unparalleled scarcity: our moon eclipsing the sun, while the sun itself is partially eclipsed by Venus. Alas, the hundreds of observers waiting around the globe to time the moment of ingress would have been frustrated by the sudden disappearance of the sun—not by a passing cloud, but by our own satellite.

On the other hand, the disappearance of a planet *behind* the sun—an occultation—is not quite as rare: one occurred on June 13, 1992; another is scheduled for June 11, 2000. Such *anti-transits* are more frequent than ordinary transits. A transit can occur only when Venus is inside an imaginary cone whose vertex is at the earth's center, and whose base is the solar disk; an anti-transit occurs when Venus is inside the *extension* of this cone on the opposite side of the sun. Since this imaginary cone opens up as we go farther from Earth's center, Venus has a greater chance of entering the extended region beyond the sun than the region in front of it. The next anti-transit of Venus will be on June 11, 2000, followed regularly at eight-year intervals until 2048, after which Venus will just miss the solar disk. This present-day cycle began in 1976.[5] Of course, because Venus is occulted *by* the sun, an anti-transit is entirely unobservable; but we may console ourselves with the thought that it is a perfectly mutual event: at the very moment when Venus disappears behind the sun in our skies, Earth will do the same in the Venusian sky. An intriguing mind-exercise it is; to quote Meeus, "We made these calculations for our own pleasure and to satisfy our curiosity—and that of the reader."[6]

Notes and Sources

1. A full account of this episode can be found in Owen Gingerich, *The Great Copernicus Chase and Other Adventures in Astronomical History* (Cambridge, Mass.: Sky Publishing Company, and Cambridge, U.K.: Cambridge University Press, 1992), chapter 19.

2. This and the subsequent information is based on Jean Meeus, *Transits* (Richmond, Va.: Willmann-Bell, 1989), pp. 31–39.

3. These excerpts are from Arthur C. Clarke, *The Wind from the Sun: Stories of the Space Age* (New York: Harcourt Brace Jovanovich, 1972), pp. 131–145. Reprinted by permission of Harcourt, Inc.

4. For details of various transits as seen from Pluto, see Jean Meeus, *More Mathematical Astronomy Morsels* (Richmond, Va.: Willmann-Bell, 2002), chapter 52.

5. See the article "Venus Hides Out" by George Lovi, *Sky & Telescope*, August 1991.

6. Jean Meeus, *Mathematical Astronomy Morsels* (Richmond, Va.: Willmann-Bell, 1997), p. 156.

Toward Distant Suns

EVEN BY astronomical standards, the news was sensational. On November 28, 2001, the *New York Times* announced that for the first time astronomers had detected, and conducted a preliminary chemical study of, the atmosphere of a planet beyond our solar system. The planet, a Jupiter-size object orbiting the 7.6 magnitude star HD 209458 in the constellation Pegasus, was discovered in 1999 and considered at first to be just one more object in a growing list of extrasolar planets discovered since 1995. By 2002, the list had grown to nearly a hundred giant planets. None, however, could be seen directly, because the faint light they receive from their parent stars and then reflect back into space is drowned by the stars' own glare. Instead, astronomers have inferred the existence of these planets from the slight gravitational pull they exert on their parent stars. This pull causes the star to appear to wobble ever so slightly in the sky, a wobble that can be detected by a sensitive spectroscopic device in much the same way as a radar detector can spot a speeding car.

Still, this kind of evidence, by its very nature, is circumstantial: though the existence of the new planet had been established beyond doubt, no one has as yet *seen* it. This was reminiscent of the events leading to the discovery of the planet Neptune, when perturbations in the orbit of Uranus led Leverrier to infer the existence of an unknown planet circling the sun beyond Uranus's orbit. In both cases the smoking gun was there, but the body was still missing. This time, however, the object being sought was some 153 light years away!

Shortly after the existence of the planet orbiting HD 209458 was established, several search teams analyzed the minute perturbations imparted to the star by its invisible planet and made a startling discovery: the planet's orbit lies almost edge on as viewed from Earth, raising the possibility that it may at times pass in front of its parent star and thus give astronomers a golden opportunity to confirm its existence directly. But unlike transits of Mercury and Venus, during which the planet's silhouette can be seen as a black dot slowly moving across the sun's face, HD 209458 is far too distant to show a visible disk, let alone a black dot crossing it. Instead, astronomers would attempt to record the slight dip in the star's brightness as the unseen planet passes in front of it.

The various search groups were able to determine that the invisible planet would transit its parent star every 3.5 days, with each transit lasting about three hours. With this data in hand, a team headed by Timothy M. Brown of the National Center for Atmospheric Research in Boulder, Colorado, and by David Charbonneau of the Harvard-Smithsonian Center for Astrophysics was able to calculate when the invisible planet would pass in front of its parent star.

In September 1999 they directed their modest 4-inch telescope, equipped with a sensitive light-detector, to HD 209458 and recorded not one but *two* transits at the precise predicted times. During each transit the star's brightness dipped by 1.6 percent (approximately 0.017 stellar magnitude) for about 2.25 hours; it took the planet an additional 30 minutes to fully enter and exit the star's invisible disk.[1] From these findings, the astronomers deduced that the transiting planet is about 30 percent larger than Jupiter, and its orbit is tilted to our line of sight by a mere 3 degrees. Even more amazingly, the planet is orbiting its parent star at a distance of just 0.05 astronomical units—about one-eighth of Mercury's distance from the sun. Conditions must surely be hellish on that remote, strange world.

But that was not all. Having established a precise timetable of the planet's transits, Charbonneau and Brown directed an imaging spectrograph aboard the Hubble Space Telescope to record the intensity of the star's light as it dimmed while the planet was partially eclipsing its parent star. The resulting spectrum revealed the existence of sodium in the planet's atmosphere—the first-ever identification of a chemical element in the atmosphere of a planet outside our solar system.[2]

For this extraordinary harvest of discoveries we must thank the near edge-on orbit of an unseen—and as yet unnamed—planet around its parent star, "a geometric gift from nature," to quote *Parade Magazine* of September 30, 2001. And this is just the beginning. Some dozen transit searches have since been undertaken, and NASA's planned Kepler mission, scheduled for launch in 2006, promises to detect many more extrasolar planets using transit photometry.[3] The day may not be far away when space telescopes will be able to detect even Earth-sized extrasolar planets and perhaps answer the age-old question: Is there life elsewhere in the universe?[4]

Notes and Sources

1. *Sky & Telescope*, February 2000, pp. 16–17. The 1.6 percent dimming of the star's brightness, small as it is, is still much larger than that of the sun during a transit of Venus: about 0.0075 percent.

In November 1999 Gregory W. Henry of Tennessee State University in Nashville, guided by data from veteran planet seekers Geoffrey W. Marcy of the University of California at Berkeley and R. Paul Butler of the Carnegie Institute in Washington, repeated the observations, this time using a robotic telescope in Arizona. See the *New York Times*, November 16, 1999, p. A21.

2. As of this writing, it was reported that the planet orbiting HD 209458 is shedding a huge stream of hydrogen, which trails behind it "like the tail of a comet" at the astounding rate of some 10,000 tons of hydrogen per second (*Sky & Telescope*, June 2003, pp. 20–21). The findings are still preliminary, but, says Charbonneau, "we don't worry too much. Transits happen twice a week, so we can always go back and observe them over and over again."

3. *Sky & Telescope*, June 2002, p. 30.

4. The *New York Times* on January 7, 2003, announced the discovery of a new extrasolar planet, OGLE-TR-56b, some 8,000 light years away, transiting its parent star every 29 hours. Unlike HD 209458, however, this planet was *first* detected by the transit method rather than by the gravitational wobble it imparts to its parent star.

A Personal Pilgrimage

IN JULY 2002 my wife and I traveled to England to visit the places associated with the 1639 transit of Venus. Our pilgrimage began at Manchester's Town Hall, a huge imposing Victorian building. Here, in the stately reception hall on the second floor, there is a set of twelve large murals, painted by the nineteenth-century artist Ford Madox Brown, depicting scenes from the history of Manchester. One mural shows William Crabtree staring in awe at the tiny black image of Venus in front of him, while his wife stands behind holding their child. Brown has obviously romanticized the event, making Crabtree appear much older than his twenty-one years at the time, but it was still moving to see the historic event memorialized on such a grand scale.

Next we traveled to Much Hoole, some 10 miles west of the town of Preston in the Lancashire district. The small, sleepy village lies amidst open fields and low hills. We walked for a while along empty streets until we saw a pedestrian walking his dog, so we stopped him and asked for directions to St. Michael's Church. We could read the puzzlement on his face: why would visitors from the United States come to this small place? So we just said we were after an astronomical event that happened here several centuries ago, upon which he said, "You mean the transit of Venus?" We soon found out that Horrocks was a household name here.

When we arrived at the church, we first paid a visit to the adjacent Hoole Church of England Primary School, where we were received with much honor. The headteacher, Mr. David Upton, introduced us to his classes. The pupils, he explained, had studied about the celestial event that took place right there over 350 years ago and were planning to watch the 2004 transit from their school yard, hoping and praying for clear skies when the moment would finally arrive early on the morning of June 8. In the hallway there was a large quilt entitled "Life in Hoole" made by the students; one square showed a telescope with Jeremiah Horrocks's name next to it.

Mr. Upton then telephoned the Rector of St. Michael's Church, Reverend Steven Hughes, who arrived within minutes on his bicycle. He kindly allowed us to tour the church, the very place where Horrocks had to attend to "business of the highest importance" on December 4, 1639, an interruption that prevented him from seeing the beginning of the transit. Several artifacts commemorate the event, and an inscription engraved in marble tells of Horrocks's brief life. Above the outside entrance to the church there is an impressive sundial with the inscription *Sine Sole Sileo* ("Without the Sun I Am Silent"). The opposite wall features a modern clock with the inscription "In memoriam Horrocii 1639–1859," erected there on the 220th anniversary of the historic transit.

We ended our trip by walking the mile or so to the Carr House, which, we learned, was up for sale (obviously this was big news in the village, as it must have been the most expensive piece of real estate in town). The house, an impressive brick structure with a beautiful garden, was closed to visitors, dashing our hopes to be allowed in by posing as potential buyers. We had to make do with admiring it from the outside and letting our imagination do the rest. A plaque at the entrance briefly tells of the event in whose wake we had come.

We left deeply moved, being transported back in time to that wintry day in 1639 when two young friends were privileged to be the first humans to watch one of astronomy's rarest events.

The author with the mural by Ford Madox Brown.

The quilt "Life in Hoole."

St. Michael's Church.

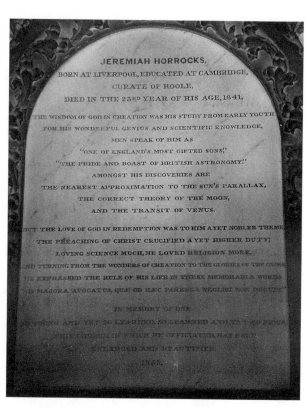

Inscription in memoriam of Jeremiah Horrocks.

Sundial with the inscription *Sine Sole Sileo* ("Without the Sun I Am Silent").

Clock in memoriam of Horrocks.

The Carr House, from whose third floor Horrocks observed the transit.

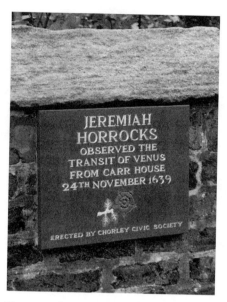

Plaque at the entrance to the Carr House.
The date is by the Old Style (Julian) calendar, then still in use in England.

12

June 8, 2004

> In passage across the face of the sun Venus is a striking sight, a black disk large enough to be detected even with the suitably protected naked eye. For watchers in 2004 there will be the awesome thought that not a single human being remains alive who observed the last transit of Venus, in December, 1882.
>
> Joseph Ashbrook, *The Astronomical Scrapbook* (1984)

WHEN VENUS returns to the sun's face on June 8, 2004, the world she saw on her last visit will be vastly changed. Mighty empires have since risen and fallen. World War I—"the war to end all wars"—was fought at an enormous human sacrifice. Yet scarcely twenty years later, the world plunged into an even mightier struggle, in whose aftermath we still live. Science, spurred by the two titanic clashes, progressed in leaps and bounds. The atom's nucleus was smashed, its binding energy unleashed with unprecedented furor. Gravity—the force that shackled humans to this planet since the beginning of time—was overcome in 1957, when the Soviet Union launched *Sputnik I.* Twelve astronauts have walked on the moon, and scores of unmanned spacecraft have visited the remotest corners of the solar system. The medium of light, until 1940 the only messenger available to us to explore the universe, was extended to include the entire electromagnetic spectrum. And observatories are no longer confined to mountaintops on earth: a dozen of them are scanning the universe from the vantage of deep space. Had a nineteenth-century astronomer been allowed to be with us today, he would have barely recognized the universe we have come to know.

Venus was not left behind. Since the 1950s, her secrets have slowly been uncovered, and it left astronomers dumbfounded. Instead of a warm, humid world, lush with flora and fauna, they discovered a desolate place, a greenhouse world with a surface temperature of 600°F—high enough to melt lead. Its thick atmosphere was found to consist of carbon dioxide and sulfuric acid, hardly a makeup conducive to life. And in 1965, radar signals transmitted to Venus from the huge radio telescope in Arecibo, Puerto Rico, finally solved the mystery of the planet's rotation period: it was found to be 243 days, the longest period of any planet in the solar system; moreover, Venus turns on its axis in a *retrograde*, east-to-west direction, instead of the west-to-east rotation of most other planets.[1] Soon thereafter, unmanned spacecraft began to visit the planet. *Mariner 2* completed the first flyby in 1962, to be followed in 1970 by *Venera 7*, the first spacecraft to land on Venus's surface. In 1990 *Magellan* went into orbit around Venus, mapping it extensively from pole to pole with radar beams that could penetrate its heavy cloud cover. Our "sister planet" was finally giving up her secrets. Far from being our twin, the Venus we know today is the least earthlike among the four "terrestrial" planets.

All this, of course, will not lessen the excitement as we approach Venus's next appointment with the sun. To be sure, scientific interest in the event will have lost much of its urgency. Since the last transit in 1882, the dimensions of the solar system have been determined with unprecedented accuracy: the distance to the moon is now known to the nearest inch, and the distances to the planets to the nearest mile, thanks to radar and laser technology. This time Venus will be greeted not by heavily funded, government-sponsored expeditions, but by enthusiastic amateurs eager to get a rare glimpse of her outline starkly etched on the solar disk.

★★★

THE transit of Tuesday, June 8, 2004, will start at 13 minutes and 27 seconds past 5 A.M. UT (Universal Time, the astronomical equivalent of Greenwich Mean Time), when a tiny notch will slowly make itself visible on the sun's southeast limb (as seen from the center of the earth); this marks the beginning of ingress. Nineteen minutes later, at 5:32:53, the planet's disk will have fully entered into the sun. This moment—second contact—will be awaited with great anticipation; it was at this instant that the famous black drop effect had first been seen. For the next five hours and thirty-three minutes, Venus's black disk will slowly move across the sun, until it reaches the sun's southwest limb at 11:06:31 (fig. 12.1). This is the moment of third contact, and again all eyes will be strained to see the black drop. Nineteen minutes later, at 11:25:57, Venus will have left the sun, not to return until June 6, 2012.[2]

The times given above are *geocentric* times, calculated for an imaginary observer at the center of the earth. But since we live on the surface of our planet, the actual times for any given location may differ from the geocentric times by up to about seven minutes. In Jerusalem, for example,

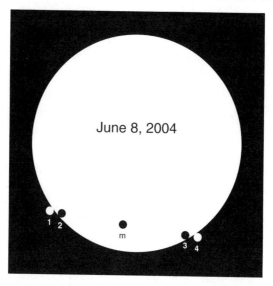

FIGURE 12.1 Passage of Venus in front of the sun,
June 8, 2004, as seen from the center of the earth.
"m" stands for mid-transit.

the transit will begin at 5:19:28 UT, that is, six minutes and
one second later than the geocentric time. To adjust Universal Time to your *local* time, you need to know by how
many hours your time zone is ahead or behind UT. Bear in
mind that the transit will take place during summer in the
Northern Hemisphere, and many countries will be on *daylight saving time* (or simply "summer time"), which is one
hour ahead of standard time. Thus in Jerusalem, two time
zones ahead of UT, the transit will begin at 8:19:28 A.M.
local summer time.

During the six hours and twelve minutes that Venus will be
moving across the sun, our own planet will have completed
about one-fourth of a revolution around its axis. Consequently,
some locations will see the entire transit from beginning to

end, while others will see only the beginning or end phases. New York, for example, will miss the beginning, which occurs during nighttime there; when the sun finally rises at 5:24 A.M. Eastern Daylight Time, Venus will have already moved more than halfway across the sun's face. But the Big Apple will at least witness the end phases of the event, with fourth contact occurring at 7:25:50 A.M. EDT. The central and western parts of the American continent will be less fortunate: the entire transit will be invisible from there. Figure 12.2 shows the earth as it would appear to an observer on the sun at the beginning and end of the transit. A table of local ingress and egress times for several major world cities is given in Appendix 2. For updated information on the circumstances of the transit at various locations around the globe, the reader should consult leading astronomical journals such as *Sky & Telescope* and *Astronomy*. Fred Espenak, of NASA's Goddard Space Flight Center, who has been issuing detailed eclipse bulletins regularly since 1991, will have details of the transit posted on his web page (http://sunearth.gsfc.nasa.gov/eclipse/transit/transit.html).

✱✱✱

UNLIKE Mercury, the image of Venus on the sun is large enough to be visible as a disk to the unaided eye. I hasten, however, to add the following:

Warning. The entire transit should be viewed only with an appropriate solar filter. Watching it with the unaided eye—even if only for a short glimpse— may result in permanent eye damage, possibly even blindness. If you use a telescope, an appropriate filter should be placed on the side of the tele-

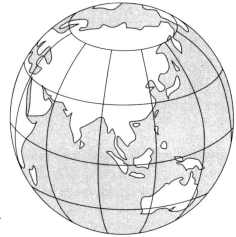

05:13:27 UT

June 8, 2004

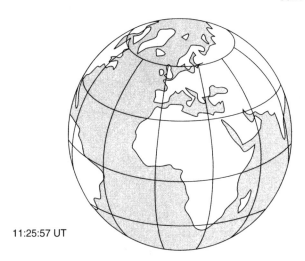

11:25:57 UT

FIGURE 12.2 Earth at the beginning and end of the transit of June 8, 2004, as seen from the sun.

scope *facing the sun*, not the side facing your eye. This will cause most of the sunlight to be filtered out *before* entering your telescope. Never look at the sun directly through your telescope, not even through your finder scope. It is advisable to cover the finder *ahead of the transit*, so as to avoid looking through it accidentally.

By "appropriate filter" I mean a solar filter sold by a reputable outlet of astronomical equipment. **Sunglasses, smoked glass, or the negative of a film are *not* regarded as safe and should not be used.** A safe filter is a #14 welder's glass, available at hardware stores. The safest method of all is to project the sun's image through a small pinhole cut in the cover of a shoe box. The sun's image, with Venus appearing as a dark dot, will be clearly visible on the side opposite of the hole. This is the method known as *camera obscura*—the same that Horrocks employed when he observed the 1639 transit. Generally speaking, the same precautions should be taken as when watching or photographing the partial phases of a solar eclipse. For more details, consult any book on viewing and photographing solar eclipses.[3]

There are a few additional items to bear in mind as you plan to watch the event. As already mentioned in chapter 5, a transit of Mercury or Venus always starts at the eastern limb of the sun and ends at the western limb; this is in contrast to solar and lunar eclipses, which usually proceed in a west-to-east direction.[4] Also, unlike a total solar eclipse, during which the excitement steadily builds up until the moment of totality but then quickly fades, here it is the beginning and end of the transit—the moments of ingress and egress—that will get all the attention. Between these moments there will be more than five hours with little else to do other than follow Venus's black silhouette as it slowly makes its way across the solar disk. So you may want to plan some activities to keep yourself busy during those five hours. If you have a telescope—again, properly equipped with a solar filter—you might take a time sequence of photographs every 30 minutes or so that will show Venus's progress across the solar disk as a series of black dots.

Unlike a total eclipse of the sun, which can only be observed from a narrow path where the moon's umbra sweeps over the earth, the entire transit will be visible from much of the Eastern Hemisphere (see again fig. 12.2). The choice of location from which to observe will therefore be dictated as much by political considerations as by the weather predictions for the area. Bear in mind that for locations in the Northern Hemisphere, this will be the beginning of summer, and quite high temperatures are possible, particularly in the Middle East and southeast Asia. Be sure to have a broad hat and sunscreen to protect against the sun's ultraviolet rays. A water bottle and some snacks are also advisable.

✱✱✱

THE years 2003–2004 happen to be exceptionally rich in astronomical events: four total lunar eclipses in a row (two in each year), a transit of Mercury on May 7, 2003, and the arrival of NASA's *Cassini* spacecraft at Saturn, starting the first closeup exploration of the ringed planet and its large moon, Titan. On August 27, 2003, the planet Mars came to within 34.6 million miles of Earth, its closest approach in nearly

60,000 years. On March 28, 2004, three of Jupiter's four large moons will be seen casting their shadows on the giant planet. And on September 10, 2004, Mercury will come within a mere 3 arc minutes of the bright star Regulus in the constellation Leo, causing the two celestial objects to appear as a double star.[5] But topping them all will be the transit of Venus on June 8.

Many of those who witnessed the five past transits of our "twin planet" have left us their impressions in words ranging from a cold, rational recording of data to an outpour of emotions. The rarity of the event made many observers conscious of the shortness of our own lifetime. An octogenarian might live long enough to witness two returns of Halley's comet; but no one among the many scientists who watched the last transit in 1882 will be around to see the next one. So let me conclude with the words Richard Anthony Proctor used in 1870 at the end of his classic book, *Transits of Venus:*

> The astronomers of the first years of the twenty-first century, looking back over the long transitless period which will then have passed, will understand the anxiety of astronomers in our own time to utilise to the full whatever opportunities the coming transits may afford; and I venture to hope that should there then be found, among old volumes on their book-stalls, the essays and charts by which I have endeavoured to aid in securing that end (perhaps even this little book in which I record the history of the matter), they will not be disposed to judge overharshly what some in our own day may have regarded as an excess of zeal.

Notes and Sources

1. The 243-day period is Venus's *sidereal* period—its rate of rotation relative to the fixed stars. Its rate of rotation relative to the *sun*—the length of the Venusian day—is 116.8 Earth days. Consequently, an observer on Venus would see the sun rising in the west and setting in the east 59 Earth days later.

2. These times are based on Jean Meeus, *Transits* (Richmond, Va.: Willmann-Bell, 1989), p. 48. They have been corrected to take into account the quantity ΔT, a correction factor due to the retardation in Earth's rotation rate, estimated at present to be sixty-seven seconds.

3. For an overall evaluation of filters, see the article "Solar Filter Safety" by B. Ralph Chou, *Sky & Telescope,* February 1998, pp. 36–40.

4. A rare exception was the recent annular eclipse of May 31, 2003, which proceeded from east to west as seen from Iceland and Greenland. This is because the moon's central shadow cone reached these locations from "above" the North Pole as viewed from space. See Jean Meeus, *More Mathematical Astronomy Morsels* (Richmond, Va.: Willmann-Bell, 2002), pp. 63–65.

5. These events are taken from Jean Meeus, *Mathematical Astronomy Morsels* (Richmond, Va.: Willmann-Bell, 1997), pp. 160 and 280.

APPENDIX 1

Halley's Method

We give here a simplified version of Halley's method for using the transit of Venus to find the Earth-Sun distance—the astronomical unit.[1]

Figure A.1 shows a cross section of the ecliptic—the plane of Earth's orbit around the sun. Because Venus's orbit is slightly tilted to this plane, the planet, during most of its path around the sun, will be above or below the ecliptic. A transit can occur only when the sun, Venus, and Earth form nearly a straight line, that is, when Venus—represented by V in the figure—is just crossing the ecliptic. (If the three bodies were *exactly* lined up, the transit would be central, but this has never happened in the past two thousand years and will not happen in the next four thousand years.)[2]

Imagine now that two observers T_1 and T_2, stationed as far apart as possible on the earth, are watching Venus during a transit; they see Venus's image on the sun at points V_1 and V_2, respectively. The triangle $T_1 V T_2$ is nearly isosceles, so if we know the length of its base b and the angle α at the vertex, we can find its height d, the distance between Earth and Venus at the time of the transit. And once d is known, the distance between the earth and the *sun*—and indeed all other solar-system distances—can be found from Kepler's third law (see page 13).

FIGURE A.1 Halley's method.

To find b is relatively straightforward: we only need to know the exact positions of the two observers in terms of their longitude and latitude on the earth. But to determine the angle α (by astronomical convention, α is twice the angle of parallax of V) is a different matter: we would need to have some celestial reference scale, and we would need it during daytime, when the fixed stars that provide this scale at night are invisible. It was Halley's idea to use the sun for this purpose: during a transit the sun would serve as a remote billboard against which α could be measured. Since the sun's apparent diameter is known (about 1,900 arc seconds), all one had to do was to simultaneously record the image of Venus on the sun's face as seen from each station—that is, points V_1 and V_2—and then measure the angular distance between them as a fraction of the sun's apparent diameter.

It sounded simple, but there was a problem: how could we ensure that the two observations would indeed be simultaneous? In today's era of instant communication, the only limit to a simultaneous measurement from two different locations is the time it takes light to travel from one location to the other. But in the eighteenth century, the observers had to rely on mechanical clocks, synchronized prior to departure, and then trust that these clocks would keep their time several months later, and at locations as far apart as the arctic circle is from the equator. This made any direct measurement of the parallax angle nearly impossible. Halley thought of a way around this problem: instead of measuring the angle directly, let each observer record the *duration* of the transit—the elapsed time between ingress and egress (more precisely, between second and third contacts). Because each observer will see Venus cross the sun along a slightly different path, these durations will be different—by up to about seventeen minutes for two observers at opposite sides on the earth (see fig. 6.1 on page 72). From the durations and Venus's known rate of motion along its orbit, one can find the length of each path as a fraction of the sun's apparent diameter. And since the two paths form parallel chords in a circle with a known diameter, one can compute the angular distance between them, the angle α.

This, then, was the essence of Halley's plan. It sounded simple, but complications arose. Aside from the uncertainties of the weather at the various observation sights, the method relied on an exact determination of the length of the baseline b, which in turn depended on an exact knowledge of the geographical longitude and latitude of the observers. The first of these parameters, latitude, was easy to determine—one only had to measure the sun's altitude above the horizon at local noon. But longitude was a different matter, and its exact determination was, until the eighteenth century, one of the most pressing problems of the maritime community.[3] To complicate things even further, triangle $T_1 V T_2$ is *nearly*, but not *exactly*, isosceles (its shape depends on the exact positions of the two observers), making the necessary computations much more difficult. But what ultimately doomed Halley's method was a problem that no one had anticipated: Venus's atmosphere. The infamous black drop effect—an apparent ligament that seemed to connect the image of Venus to the sun's limb for several seconds immediately after second contact and again just before third contact—made any precise timing of the transit nearly impossible.

Toward the second half of the nineteenth century, alternative methods for determining the astronomical unit were developed—in particular, using the planet Mars and some of the minor planets (asteroids) for parallax measurements. These techniques made Halley's method obsolete, even as expeditions were sent around the world to observe the 1874 and 1882 transits. The ultimate determination of the dimensions of our solar system had to wait until the development of radio astronomy. On April 10, 1961, a radar beam was transmitted to Venus from the Jet Propulsion Laboratory in California. By recording the time it took the beam to reach Venus, bounce off its surface and return to Earth, the distance between the two planets was found to be 26,372,600 miles, with an uncertainty of only a few hundred miles. This single determination was sufficient to fix the scale of the solar system with an unprecedented accuracy.

Notes and Sources

1. Excerpts from Halley's paper of 1716 in which he proposed his method can be found in Harlow Shapley and Helen E. Howarth, *A Source Book in Astronomy* (New York: McGraw-Hill, 1929), pp. 96–100.

2. The closest a transit came to be central was on November 23, A.D. 424, when Venus passed just 9.6 arc seconds above the sun's center.

3. See Dava Sobel, *Longitude* (New York: Walker and Co., 1995).

APPENDIX 2

Times of the Transit of June 8, 2004,
for Some Major Cities

City	Ingress t_1	t_2	Midtransit t_m	Egress t_3	t_4
Beijing	05:13:10	05:32:21	08:15:00	10:59:22	11:18:58
Berlin	05:19:40	05:39:27	08:22:04	11:03:25	11:22:53
Bombay	05:16:13	05:35:11	08:18:38	11:02:31	11:21:36
Buenos Aires	—	—	—	11:13:34	11:32:48
Chicago	—	—	—	11:05:14	11:25:18
Jerusalem	05:19:28	05:38:52	08:22:02	11:04:07	11:23:16
London	05:19:52	05:39:45	08:22:40	11:04:03	11:23:34
Moscow	05:18:48	05:38:26	08:20:34	11:02:04	11:21:32
New York	—	—	—	11:05:52	11:25:50
Paris	05:19:59	05:39:51	08:22:50	11:04:14	11:23:43
Rio de Janeiro	—	—	—	11:13:04	11:32:14
Rome	05:20:09	05:39:53	08:22:59	11:04:31	11:23:51
Sydney	05:07:20	05:26:07	—	—	—
Teheran	05:18:25	05:37:43	08:20:26	11:02:42	11:21:54
Tokyo	05:11:12	05:30:26	08:13:44	—	—
Toronto	—	—	—	11:05:19	11:25:21

Notes and Sources: I am indebted to Fred Espenak, of NASA's Goddard Space Flight Center, for supplying me with these figures. All times in this table are Universal Times (UT), the astronomical equivalent of Greenwich Mean Time. t_1 and t_2 are the contact times at *ingress*, the moments when Venus first enters onto the sun's disk and when it has fully entered onto it, respectively. t_3 and t_4 are the corresponding times at *egress*, when Venus is about to leave the sun. t_m is the time of midtransit, the moment of closest approach to the sun's center.

APPENDIX 3

Dates of Some Past and Future Transits

The data in the following five tables is based on Jean Meeus, *Transits* (Richmond, Va.: Willmann-Bell, 1989), pp. 31–49. This publication also has information about transits visible from the outer planets—Jupiter, Saturn, Uranus, and Neptune.

TABLE 1. Transits of Mercury 1900–2100

1907 Nov 14	1957 May 6	2003 May 7	2052 Nov 9
1914 Nov 7	1960 Nov 7	2006 Nov 8	2062 May 10
1924 May 8	1970 May 9	2016 May 9	2065 Nov 11
1927 Nov 10	1973 Nov 10	2019 Nov 11	2078 Nov 14
1937 May 11	1986 Nov 13	2032 Nov 13	2085 Nov 7
1940 Nov 11	1993 Nov 6	2039 Nov 7	2095 May 8
1953 Nov 14	1999 Nov 15	2049 May 7	2098 Nov 10

TABLE 2. Transits of Venus 1500–2500

1518 May 26	1874 Dec 9	2247 Jun 11
1526 May 23	1882 Dec 6	2255 Jun 9
1631 Dec 7	2004 Jun 8	2360 Dec 13
1639 Dec 4	2012 Jun 6	2368 Dec 10
1761 Jun 6	2117 Dec 11	2490 Jun 12
1769 Jun 3	2125 Dec 8	2498 Jun 10

For the benefit of astronauts who might roam the inner solar system in the next century, we list in tables 3 and 4 the dates of recent and coming transits visible from Venus and Mars.

TABLE 3. Transits of Mercury Visible from Venus

1950 Oct 28	1971 Jun 11	2012 Dec 18
1954 Oct 27	1976 Dec 25	2016 Dec 17
1956 May 13	2005 Nov 17	2022 Jul 2
1960 May 12	2007 Jun 4	2028 Jan 16
1965 Nov 25	2011 Jun 3	2033 Aug 1

TABLE 4. Transits of Venus and Earth Visible from Mars

Transits of Venus		*Transits of Earth*	
1900 Oct 24	2030 Aug 20	1516 Oct 27[a]	1984 May 11[b]
1928 Mar 12	2032 Jun 18	1595 Nov 10	2084 Nov 10
1932 Oct 23	2059 Nov 5	1621 May 5	2163 Nov 15
1964 Oct 22	2064 Jun 17	1700 May 8	2189 May 10
1966 Aug 22	2091 Nov 5	1800 Nov 9	2268 May 13
1994 Jan 7	2096 Jun 16	1879 Nov 12	2368 Nov 13
1998 Aug 21	2098 Apr 16	1905 May 8	2394 May 10

[a]By the Julian calendar, which was in use until 1582.

[b]This transit inspired Arthur C. Clarke to write his story "Transit of Earth" (see p. 152).

For those planning to watch any of these events, the data in table 5 might be useful.

TABLE 5. Additional Data about Transits of the Inner Planets

Transit as Seen from	Mean Diameter of Sun (arc seconds)	Synodic Period of Transiting Planet (days)		Mean Diameter of Transiting Planet (arc seconds)	Maximum Duration of Transit (hours)
Venus	2653	Mercury	145	20	6.2
Earth	1919	Mercury	116	11	6.5
		Venus	584	61	7.9
Mars	1260	Mercury	101	6	6.9
		Venus	334	21	8.6
		Earth	780	33	9.5

BIBLIOGRAPHY

Ashbrook, Joseph. *The Astronomical Scrapbook: Skywatchers, Pioneers, and Seekers in Astronomy,* chapters 42–45. Cambridge, Mass.: Sky Publishing Corporation, and Cambridge, U.K.: Cambridge University Press, 1984.

Ball, Sir Robert S. *The Story of the Heavens,* chapter 8. London: Cassell and Company, 1910 [1886].

Baum, Richard, and Sheehan, William. *In Search of Planet Vulcan: The Ghost in Newton's Clockwork Universe.* New York and London: Plenum Trade, 1997.

Berry, Arthur. *A Short History of Astronomy from Earliest Times through the Nineteenth Century.* 1898; rpt. New York: Dover, 1961.

Brickel, Robert. *The Transits of Venus, 1639–1874, or, A Chapter of Romance in Science: In Memoriam of Horroccii.* Hoole, U.K.: St. Michael's Church, 1998.

Caspar, Max. *Kepler.* Trans. C. Doris Hellman. 1959; rpt. New York: Dover, 1993.

Cattermole, Peter, and Moore, Patrick. *Atlas of Venus.* Cambridge, U.K.: Cambridge University Press, 1997.

Chapman, Allan. *Three North Country Astronomers,* part 1. Swinton, Manchester, U.K.: Neil Richardson, 1982.

Clerke, Agnes M. *A Popular History of Astronomy during the Nineteenth Century,* part 2, chapter 6. 1885; 3d ed. London: Adams and Charles Black, 1893.

Cook, Alan. *Edmond Halley: Charting the Heavens and the Seas.* Oxford: Clarendon Press, 1998.

Danson, Edwin. *Drawing the Line: How Mason and Dixon Surveyed the Most Famous Border in America,* chapters 5 and 20. New York: John Wiley, 2001.

Gillispie, Charles Coulston, ed. *Dictionary of Scientific Biography.* 16 vols. New York: Charles Scribner's Sons, 1970–1980.

Gingerich, Owen. *The Great Copernicus Chase and Other Adventures in Astronomical History,* chapter 15 ("Johannes Kepler and the Rudolphine Tables"). Cambridge, Mass.: Sky Publishing Corporation, and Cambridge, U.K.: Cambridge University Press, 1992.

Grinspoon, David Harry. *Venus Revealed.* New York: Addison-Wesley, 1997.

Helden, Albert Van. *Measuring the Universe: Cosmic Dimensions from Aristarchus to Halley.* Chicago and London: University of Chicago Press, 1985.

Hindle, Brooke. *The Pursuit of Science in Revolutionary America, 1735–1789,* chapter 8. Chapel Hill, N.C.: University of North Carolina Press, 1956.

Hirshfeld, Alan W. *Parallax: The Race to Measure the Cosmos.* New York: W. H. Freeman, 2001.

Horrox [Horrocks], Jeremiah. *The Transit of Venus Across the Sun, to which is prefixed A Memoir of his Life and Labours.* Trans. Rev. Arundell Blount Whatton. London: William Macintosh, 1859.

Hough, Richard. *Captain James Cook: A Biography,* chapters 4, 7, and 8. New York: W. W. Norton, 1994.

Koestler, Arthur. *The Watershed: A Biography of Johannes Kepler.* Garden City, N.Y.: Anchor Books, Doubleday, 1960.

Kuhn, Thomas S. *The Copernican Revolution: Planetary Astronomy in the Development of Western Thought.* Cambridge, Mass., and London: Harvard University Press, 1985.

Ley, Willy. *Watchers of the Skies: An Informal History of Astronomy from Babylon to the Space Age,* chapter 8. New York: Viking, 1966.

Meeus, Jean. *Transits.* Richmond, Va.: Willmann-Bell, 1989.

Meeus, Jean. *Mathematical Astronomy Morsels.* Richmond, Va.: Willmann-Bell, 1997.

Meeus, Jean. *More Mathematical Astronomy Morsels,* chapters 48 and 52. Richmond, Va.: Willmann-Bell, 2002.

Mendoza, E., ed. *A Random Walk in Science: An Anthology Compiled by R. L. Weber,* pp. 174–76 ("The Transit of Venus," abridged from the English translation of Horrocks's *Venus in sole visa*). London and Bristol: The Institute of Physics, 1973.

Menshutkin, Boris N. *Russia's Lomonosov.* Princeton, N.J.: Princeton University Press, 1959.

Moore, Patrick. *Venus*. London: Cassell Illustrated, 2002.

Moore, Patrick, and Maunder, Michael. *Transits: When Planets Cross the Sun*. New York: Springer Verlag, 2000.

Newcomb, Simon. *Popular Astronomy*, part 2, chapter 3. New York: Harper, 1880.

Newcomb, Simon. *Reminiscences of an Astronomer*, chapter 6. Boston and New York: Houghton, Mifflin, and Co., 1903.

Newcomb, Simon, and Holden, Edward S. *Astronomy*, part 1, chapter 10, and part 2, chapter 3. New York: Henry Holt and Company, 1887.

Nunis, Doyce B., Jr., ed. *The 1769 Transit of Venus: The Baja California Observations of Jean-Baptiste Chappe d'Auteroche, Vincente de Doz, and Joaquin Velázquez Cardenas de León*. Translations by James Donahue, Maynard J. Geiger, and Iris Wilson Engstrand. Los Angeles: Natural History Museum of Los Angeles County, 1982.

Pannekoek, A. *A History of Astronomy*, chapters 28 and 29. 1961; rpt. New York: Dover, 1989.

Proctor, Mary. *Romance of the Sun*, chapters 2, 3, and 4. New York and London, 1927.

Proctor, Richard A. *Transits of Venus: A Popular Account of Past and Coming Transits from the First Observed by Horrocks A.D. 1639 to the Transit of A.D. 2012*. 1874; 4th ed. London: Longmans, Green, and Co., 1882.

Ronan, Colin A. *Edmond Halley: Genius in Eclipse*. London: Macdonald, 1969.

Sellers, David. *The Transit of Venus: The Quest to Find the True Distance of the Sun*. Leeds, U.K.: Magavelda Press, 2001.

Shapley, Harlow, and Howarth, Helen E. *A Source Book in Astronomy*, pp. 58–62 and 96–100. New York: McGraw-Hill, 1929.

Sheehan, William. *Worlds in the Sky: Planetary Discovery from Earliest Times through Voyager and Magellan*. Tucson: University of Arizona Press, 1992.

Williams, Pearce L. *Album of Science: The Nineteenth Century*, pp. 84–95. New York: Charles Scribner's Sons, 1978.

Woolf, Harry. *The Transits of Venus: A Study of Eighteenth-Century Science*. Princeton, N.J.: Princeton University Press, 1959.

Young, Charles A. *A Text-Book of General Astronomy for Colleges and Scientific Schools*, chapter 17. 1888; rev. ed. Boston: Ginn and Company, 1904.

Websites

The past few years have seen an enormous proliferation of websites dedicated to eclipses and transits. As of this writing, the search engine Google lists over 16,000 items under the heading "Transits of Venus." It would be next to impossible to list even a fraction of them. However, the following website gives authoritative and detailed information on the upcoming transits of 2004 and 2012:

http://sunearth.gsfc.nasa.gov/eclipse/transit/venus0412.html

ILLUSTRATION CREDITS

Figures 4.1, 5.1–5.5, 6.1, 7.1, 8.2, 11.1, 11.2, 12.1, A.1, and figs. 1–3 in the "Solar and Stellar Parallax" sidebar drawn by Eyal Maor.

Figure 1.1 Frontispiece of the *Rudolphine Tables.* From Johannes Kepler, *Tabulae Rudolphinae,* 1627. (Reprinted from Owen Gingerich, *The Great Copernicus Chase.* Cambridge, Mass.: Sky Publishing, 1992.)

Figure 1.2 Kepler's model of the universe. From Johannes Kepler, *Mysterium cosmographicum,* 1597. (Reprinted from Arthur Koestler, *The Watershed.* New York: Doubleday Anchor, 1960.)

Figure 2.1 The Copernican system. From *De revolutionibus orbium coelestium,* 1543.

Figure 2.2 Epicycles. From the author's collection.

Figure 5.4 Eleven transits of Venus, 1396–2012, adapted from Jean Meeus, *Transits* (Richmond, Va.: Willmann-Bell, 1989), with permission.

Figure 7.2 A mappemonde of the 1761 transit. (Reprinted from Harry Woolf, *The Transits of Venus.* Princeton: Princeton University Press, 1959.)

Figure 7.3 Halo around Venus just before ingress. (Reprinted from Camille N. Flammarion, *Popular Astronomy.* St. Petersburg, 1900, in Russian translation.)

Figure 8.1 The black drop effect. (Reprinted from Simon Newcomb, *Popular Astronomy.* New York: Harper and Brothers, 1880).

Figure 8.3 The effect of atmospheric distortion. (Reprinted from Richard A. Proctor, *Transits of Venus.* London: Longmans, Green and Co., 4th ed., 1882.)

Figure 9.1 *Daily Graphic* article on the Irish Expedition of 1874. From *The Daily Graphic* (New York), July 9, 1874.

Figure 9.2 Artificial transit. (Reprinted from Newcomb, *Popular Astronomy.*)

INDEX